Advances in Intelligent and Soft Computing 76

Editor-in-Chief: J. Kacprzyk

Advances in Intelligent and Soft Computing

Further volumes of this series can be found on our homepage: springer.com

Vol. 62. B. Cao,
T.-F. Li, C.-Y. Zhang (Eds.)
Fuzzy Information and Engineering Volume 2, 2009
ISBN 978-3-642-03663-7

Vol. 63. Á. Herrero, P. Gastaldo,
R. Zunino, E. Corchado (Eds.)
Computational Intelligence in Security for Information Systems, 2009
ISBN 978-3-642-04090-0

Vol. 64. E. Tkacz, A. Kapczynski (Eds.)
Internet – Technical Development and Applications, 2009
ISBN 978-3-642-05018-3

Vol. 65. E. Kącki, M. Rudnicki,
J. Stempczyńska (Eds.)
Computers in Medical Activity, 2009
ISBN 978-3-642-04461-8

Vol. 66. G.Q. Huang,
K.L. Mak, P.G. Maropoulos (Eds.)
Proceedings of the 6th CIRP-Sponsored International Conference on Digital Enterprise Technology, 2009
ISBN 978-3-642-10429-9

Vol. 67. V. Snášel, P.S. Szczepaniak,
A. Abraham, J. Kacprzyk (Eds.)
Advances in Intelligent Web Mastering - 2, 2010
ISBN 978-3-642-10686-6

Vol. 68. V.-N. Huynh, Y. Nakamori,
J. Lawry, M. Inuiguchi (Eds.)
Integrated Uncertainty Management and Applications, 2010
ISBN 978-3-642-11959-0

Vol. 69. E. Piętka and J. Kawa (Eds.)
Information Technologies in Biomedicine, 2010
ISBN 978-3-642-13104-2

Vol. 70. Y. Demazeau, F. Dignum,
J.M. Corchado, J. Bajo Pérez (Eds.)
Advances in Practical Applications of Agents and Multiagent Systems, 2010
ISBN 978-3-642-12383-2

Vol. 71. Y. Demazeau, F. Dignum,
J.M. Corchado, J. Bajo, R. Corchuelo,
E. Corchado, F. Fernández-Riverola,
V.J. Julián, P. Pawlewski, A. Campbell (Eds.)
Trends in Practical Applications of Agents and Multiagent Systems, 2010
ISBN 978-3-642-12432-7

Vol. 72. J.C. Augusto, J.M. Corchado,
P. Novais, C. Analide (Eds.)
Ambient Intelligence and Future Trends, 2010
ISBN 978-3-642-13267-4

Vol. 73. J.M. Corchado, P. Novais,
C. Analide, J. Sedano (Eds.)
Soft Computing Models in Industrial and Environmental Applications, 5th International Workshop (SOCO 2010), 2010
ISBN 978-3-642-13160-8

Vol. 74. M.P. Rocha, F.F. Riverola,
H. Shatkay, J.M. Corchado (Eds.)
Advances in Bioinformatics, 2010
ISBN 978-3-642-13213-1

Vol. 75. X.Z. Gao, A. Gaspar-Cunha,
M. Köppen, G. Schaefer,
J. Wang (Eds.)
Soft Computing in Industrial Applications, 2010
ISBN 978-3-642-11281-2

Vol. 76. T.J. Bastiaens, U. Baumöl,
B.J. Krämer (Eds.)
On Collective Intelligence, 2010
ISBN 978-3-642-14480-6

Theo J. Bastiaens, Ulrike Baumöl, and
Bernd J. Krämer (Eds.)

On Collective Intelligence

Editors

Theo J. Bastiaens
Faculty of Cultural and Social Sciences
FernUniversität in Hagen
Universitätsstrasse 11
58097 Hagen, Germany
E-mail: theo.bastiaens@fernuni-hagen.de

Ulrike Baumöl
Faculty of Economics
FernUniversität in Hagen
Universitätsstrasse 41
58097 Hagen, Germany
E-mail: ulrike.baumoel@fernuni-hagen.de

Bernd J. Krämer
Faculty of Mathematics and
Computer Science
FernUniversität in Hagen
Universitätsstrasse 27
58097 Hagen, Germany
E-mail: bernd.kraemer@
fernuni-hagen.de

ISBN 978-3-642-14480-6 e-ISBN 978-3-642-14481-3

DOI 10.1007/978-3-642-14481-3

Advances in Intelligent and Soft Computing ISSN 1867-5662

Library of Congress Control Number: 2010938115

Typeset & Cover Design: Scientific Publishing Services Pvt. Ltd., Chennai, India.

Printed on acid-free paper
5 4 3 2 1 0
springer.com

Preface

Welcome to the proceedings of the inaugural Symposium on Collective Intelligence (COLLIN 2010). This was the first of a new series of events that will evolve over the coming years, and we were happy to hold the event in Hagen where the idea for this symposium was born. The participants visited Hagen in April, with excellent opportunities to get rain, wind and sun.

Collective intelligence denotes a phenomenon according to which the purposeful interaction between individuals creates intelligent solutions and behaviors that might not have come to existence without this concerted effort of a community. The members of such communities form a social network, typically over the Internet. They are engage with each other over a sustained period of time to develop an area of innovation through collaboration and exchange of ideas, experiences and information. Leading-edge information and communication technologies (ICT) offer ample opportunities for enabling collective intelligence.

COLLIN aims to become the flagship conference in the areas collective intelligence and ICT-enabled social networking, which is attracting more and more researchers and practitioners from both academia and industry. The beginnings are extremely promising. We were delighted to receive contributions from different parts of the world including Australia, Korea and the United States. In fact, the success of an event like this depends on the quality of the papers and on the organizational efforts of the symposium officers and secretariat.

Each paper submitted was reviewed by at least two reviewers. The reviews concentrated primarily on originality, high quality and relevance to the theme of the symposium. In the end, 9 outstanding papers were accepted for presentation. The reasons for choosing so few were not only to make sure that the papers presented were of the highest quality, but, just as important, we wanted to avoid parallel session and thus facilitate interaction and exchange of ideas among participants. In addition, we invited a few renowned experts in the field to contribute to the success of this symposium with outstanding papers reporting on their most recent research.

Our special thanks go to the authors for submitting their papers to the symposium, to the international program committee, and to the numerous reviewers who did an excellent job in guaranteeing that the articles in this volume are of very high quality.

On the organization side, we are indebted to all the symposium officers for their generous, invaluable help and support in all aspects of the organization of this symposium. In particular, the local arrangements team, led by Henrik Ickler, did an outstanding job under great time pressure. We also thank Dr. Peng Han for managing the registrations, and special thanks are due to the Gesellschaft der Freunde der FernUniversität e.V. who generously sponsored the social events of this symposium.

April 2010

Theo Bastiaens
Ulrike Baumöl
Bernd Krämer

Organization

COLLIN 2010 Symposium Organization

COLLIN 2010 was jointly organized by the Faculty of Cultural and Social Sciences, the Faculty of Economics and the Faculty of Mathematics and Computer Science of FernUniversität in Hagen.

Symposium and Program Chairs

Theo J. Bastiaens	FernUniversität in Hagen, Germany
Ulrike Baumöl	FernUniversität in Hagen, Germany
Bernd J. Krämer	FernUniversität in Hagen, Germany

Organizing Chair

Henrik Ickler	FernUniversität in Hagen, Germany

Registration Chair

Peng Han	ChongQing Academy of Science and Technology, China

Program Committee

Jörn Altmann	Seoul National University, Korea
Aurélie Aurilla Arntzen Bechina	College University i Buskerud, Norway
Homa Bahrami	University of California at Berkeley, USA
Sabine Fliess	FernUniversität in Hagen, Germany
Peter A. Gloor	MIT Center for Collective Intelligence, USA
Jörg M. Haake	FernUniversität in Hagen, Germany
Axel Hochstein	Stanford University, USA
Norbert Hoffmann	Swiss Life AG, Zürich, Switzerland
Reinhard Jung	Universität St. Gallen, Switzerland
Wilhelm Rödder	FernUniversität in Hagen, Germany
Gunter Schlageter	FernUniversität in Hagen, Germany
Chen-Yu Phillip Sheu	University of California at Irvin, USA
Klaus Tochtermann	Universität Graz, Austria
Brigitte Werners	Ruhr-Universität Bochum, Germany
Claudia de Witt	FernUniversität in Hagen, Germany
Olaf Zawacki-Richter	FernUniversität in Hagen, Germany

Table of Contents

On Collective Unintelligence 1
Mark McGovern

Building Actor Reputation in Web-Based Innovation Networks 13
Sabine Fliess, Arwed Nadzeika, Marco Wehler, Jorinde Wormsbecher

An Approach for the Visual Representation of Business Models That Integrate Web-Based Collective Intelligence into Value Creation 25
Henrik Ickler

Open Science 2.0: How Research and Education Can Benefit from Open Innovation and Web 2.0 37
Oliver Tacke

A Social Network System for Analyzing Publication Activities of Researchers 49
Alireza Abbasi, Jörn Altmann

Use of Swarm Intelligence to Involve Customers in Product Innovation 63
Sandro Georgi, Reinhard Jung

Imitation and Quality of Tags in Social Bookmarking Systems – Collective Intelligence Leading to Folksonomies 75
Fabian Floeck, Johannes Putzke, Sabrina Steinfels, Kai Fischbach, Detlef Schoder

Measuring and Analyzing the Openness of the Web2.0 Service Network for Improving the Innovation Capacity of the Web2.0 System through Collective Intelligence 93
Kibae Kim, Jörn Altmann, Junseok Hwang

Collective Intelligence in Teams – Practical Approaches to Develop Transactive Memory 107
Michael W. Busch, Dietrich von der Oelsnitz

The Need Language: A Preliminary Report 121
A. Abramovich, C.Z. Xu, P. Guo, L. Wang, T. Qian, Q. Wang, P.C.-Y. Sheu

Adding Taxonomies Obtained by Content Clustering to Semantic Social Network Analysis 135
Hauke Fuehres, Kai Fischbach, Peter A. Gloor, Jonas Krauss, Stefan Nann

How to Reduce New Product Development: Customer Integration in the e-Fashion Market 147
Frank T. Piller, Evalotte Lindgens

Author Index 159

On Collective Unintelligence

Mark McGovern

School of Economics and Finance, Queensland University of Technology

Abstract. The idea of collective unintelligence is examined in this paper to highlight some of the conceptual and practical problems faced in modeling groups. Examples drawn from international crises and economics provide illustrative problems of collective failures to act in intelligent ways, despite the inputs and efforts of many skilled and intelligent parties. Choices made of "appropriate" perceptions, analysis and evaluations are examined along with how these might be combined. A simple vector representation illustrates some of the issues and creative possibilities in multi-party actions. Revealed as manifest (un-)intelligence are the resolutions of various problems and potentials that arise in dealing with the "each and all" of a group (wherein items are necessarily non-parallel and of unequal valency). Such issues challenge those seeking to model collective intelligence, but much may be learned.

Keywords: intelligence, paradoxes, crisis, resolution, international economics.

1 Introduction

Collective unintelligence is examined in this paper through considerations of illustrative problems and approaches from economics. The goals are to foster discussion of issues that appear to be important when intelligence is to be applied and to explore approaches with some apparent potential. The underlying position is that effective intelligence involves creative processes and dynamic balances which are inhibited by an overreliance upon formal or prescriptive methods.

Examples of collective unintelligence abound in our world with the ongoing global financial and economic crisis probably the most outstanding current example. Not only was this "event" unexpected by most, but importantly not all. Surprisingly little coherence is evident in recovery strategies. Worse still, the actions of many nations may well be escalating what has been a banking crisis into crises of national default and currency destruction. Two years into the acute phase of the crisis, collective processes have effectively generated a spread and deepening of problems, a commitment to risk, a potential for further failures and little real progress.

"Less unintelligence" in dealing with such problems is needed. Considerations from international economics (along with some from industrial economics and international business) are used to outline some core problems. Reflections offered draw from theoretical and applied analyses and well as observations on the educational experiences of tertiary students. Together these help sketch the nature of "failure" in what de Bono [1] termed first stage (or more perceptual) thinking. Inappropriate choices (be they of specification, positions, information, influences, focii and frames) are all potential sources of errors, failures and "unintelligence".

T.J. Bastiaens, U. Baumöl, and B.J. Krämer (Eds.): On Collective Intelligence, AISC 76, pp. 1–11.
springerlink.com

A central conceptual issue is the types of association and selection favoured in any multi-party situation. How this is then treated is particularly critical in second-stage (or more analytic) processes. An illustrative application using vector analysis and non-parallel items allows consideration of not just crises but also of the influences from prevailing methods. Intelligence is in part an exercise in avoidance of undue dominance. Brief consideration of the *aufheben* in Hegel's dialectic shows how some favoured analytics can strongly flavour perceptions, actions, outputs and outcomes.

Notionally, applied intelligence is evident when projects and projections "work out satisfactorily". While satisfaction may be "merely" in terms of survival, typically something "more" is sought. Enhancement of some or other attributes may be a goal of an intelligent agent or group. Changes in quantity and quality style development, and these may be variously evaluated, including by various group members. Successful collectives dynamically and creatively resolve differences, to the sufficient satisfaction of each and all. "Unintelligent" ones fail in one or more of these aspects.

Arguably, intelligence as manifest in skilful use of appropriate tools (including thoughts and contributory inputs) and effective achievement of desired outcomes is in short supply in significant areas of the economy. Examining "unintelligence issues" may help elucidate what needs to be done if collective gains are to be more readily achieved and distributed.

2 Formulating Collective Unintelligence

In contrast to collective intelligence which is actively researched as this conference and others attest, "collective unintelligence" is little discussed. Yet all manner of maladaptive or otherwise unintelligent behaviours are evident in economies and economics. Attendant crises abound. History is replete with booms followed by busts, of manias and panics, and of fallen mindsets and beliefs. The societal bases of crises and associated "unintelligence" provide a convenient focus.

People in an interdependent group can worsen situations by trying to do what is individually perceived to be "good" or "right". Keynes [2] provided an example in the paradox of thrift: consumers can be *worse off by saving more* due to the aggregate or collective decrease in generated consumer demand and an associated greater fall in incomes as growth and investment ease. Thus, collective savings fall despite greater individual efforts to save.

Ideas may become debased, with the use of international trade theory providing an example. "All of the things that have been painfully learned through a couple of centuries of hard thinking about and careful study of the international economy have been swept out of public discourse" by pop internationalism [3]. Such issues are no longer historical or intellectual curiosities as crisis-hit nations search for better collective outcomes and some "desperately needed" solutions.

Analytic discussions are often couched in formal terms (and may address narrow or technical aspects or such things as logical fallacies of composition). It is striking how poorly such treatments travel beyond the immediate analysis and literature, or in the minds of students. Central points often seem to be lost in a fog of confusion and detail. "Why?" questions that can be suggested include:

— why a critical or crisis situation can or does occur in a collective setting?
— why the body of relevant analysis does not disseminate more fully? and
— why students and intelligent others struggle with explanations offered?

Relaying particular or common understandings involves projecting insights into event occurrences *via* information dissemination, idea comprehension and other steps. Evaluation of any such understanding may include considerations of such things as:

— the sensitivity to definitions adopted;
— the problems of resolving many intelligences or positions; and
— the issues of relevant emphasis.

Such "how" issues while often noted are typically then worked around by assumption. Exploring these three "how" issues further shows some possibilities and problems.

1. **Definitions**. Pfeifer and Scheier [4], for example, who comment that intelligence is "hard to define" and that "not much agreement has been achieved" (p 6) find a "common denominator" in adaptive behaviour which has "two components: complying with existing rules and generating new behavior" (pp 20-21). Other definitionals could be used, such as preserving some properties other than rules. However, unchanging behavior may be adaptive, including over some horizons but not others, and may involve explicit and appropriate choices. These comments are not meant to criticise good work but rather to point out the special and sometimes still-ambiguous nature of any definitional basis. Further, when or how might definitionally intelligent behavior be situationally unintelligent? What accords with definitions need not accord with some wider reality.
2. **One or many intelligences**, and, if the latter, how these should be reconciled. Such multiplicity can occur within one party ***or*** between many.
    a. The multiple intelligences of the single human [5] have been variously expressed[1] but achieving effective balances appears under-considered. How might suitable balances be "objectively" or impersonally achieved, or are they inherently personalised?
    b. In any group of "intelligent" parties there will be a multitude of "intelligences" and any expectations of a single collective preference are bold, situationally and analytically. Cournot [7] was among the first to mathematically explore situations of economic interdependence wherein the outcomes for one (and for all) depend sensitively and in part on the actions of some other. His work and much of game theory demonstrate well many problems "of interdependence" which remain essentially unresolved today.

[1] As expressed by 6. Armstrong, T. *Multiple intelligences*. 2000 [cited; Available from: http://www.thomasarmstrong.com/multiple_intelligences.htm. the eight intelligences are: Linguistic intelligence ("word smart"); Logical-mathematical intelligence ("number/reasoning smart"); Spatial intelligence ("picture smart"); Bodily-Kinesthetic intelligence ("body smart"); Musical intelligence ("music smart"); Interpersonal intelligence ("people smart"); Intrapersonal intelligence ("self smart"); and Naturalist intelligence ("nature smart").

c. Ricardo [8] preferred the bold position, arguing for mutual gains from trade *via* comparative advantage for any nations. For some, intelligence became belief in a rule with the "can gain" of possibility became a dogmatic "must", particularly under Empire. The limitations and setting of Ricardo's work remain often ignored.
d. Olson [9] considered how tight minorities could drive collective positions, even if the majority was to be disadvantaged.

3. **Approach choice, focus and relevance**. Arguments abound about excessive attention to some preferred aspect(s) or intelligence(s) of an individual or group.
   a. The recent popularization of the "emotional intelligence" argument [10] enlivened the debate but gains made remain uncertain. Interestingly "rationality" is the preferred touchstone in economics.
   b. De Bono [11] sees potential in creative sequencing of different "hats"[2] which help direct focus and attention in thinking to achieve a more effective and constructive application of intelligences.
   c. In terms of thinking about information (within the white hat, say) de Bono [12] proposes six complementary frames[3]. We may ponder potential insights from framing under the other coloured hats.

Related is the question of whether terms of intelligent or unintelligent add anything of value, and indeed what do the terms mean? "Intelligence" remains a highly contested term and its exercise involves some fluidity in action and interpretation.

"Unintelligence" could arise from adverse influences in any of these areas. Note all these aspects are *prior to* any detailed analysis. "Something somewhere in how things were approached does or did not work out" would be a forensic expectation when approaching an incident scene of potential or realized collective unintelligence.

A particular approach is now built from a working definition of *"intelligence"* as *"an ability to project successfully"*. While there may be an interpretative bias towards a focus on *"ability"*, ***unintelligence*** might involve some inability, failure(s) in and/or of projection and/or lack of success. To explore this working basis, chosen situations and specifications will be both analysed and suggestively constructed.

This definition of intelligence can be expounded in various ways, for example as:

an ability of ___ *to project* ___ some item *successfully* ___ on some grounds

Success is now set here in terms of goals with evaluation to be somehow by comparison with them. Such a phrase can be populated in a variety of ways. Illustrative examples could include *an ability of*

---

[2] The six hats (with focus) are: white (information); red (feelings, emotions and intuition); black (faults, weaknesses, risks); yellow (values, benefits and how to achieve); green (creative effort); and blue (organization of thinking).. Arguably much discussion is white then black, a presentation of information with highlighting of weaknesses: in an informal experiment, advanced undergraduate students studying tourism were markedly reluctant to engage in hat thinking beyond the white and black. Some found the whole process most confronting.

[3] The six areas of attention (and frame) are: purpose (triangle); accuracy (circle); point of view (square); interest (heart); value (diamond); and outcome (slab).

— an organism *to project* past a danger *so as to* survive or build a relationship
— a group of workers *to project* their efforts *so as to* produce and prosper
— a baby *collecting toys* in a basket *so as to* carry more of 'mine'

Each of these could be deemed "economic" as they involve resource use. They could be alternately deemed as psychological (mind involving), physiological (involving use of a physical body), aspirational (as they may evidence purpose) and so on.

Each could be debated in terms of the nature or type of intelligence evident, and as to why any instance or example qualifies as an exemplar of intelligence at all. Does the organism, worker or baby have to be human, or organic – and if not, why? Such debates while important are beyond the scope of this paper. The approach here is to draw a little from centuries of economic investigations whereby "collective intelligence" was built *so as to* advance the interests of all and, variously, each. It was after all "the Wealth of Nations" that drew the attention of Adam Smith [13] and of many since. Much was learnt from failures, and to these we now turn.

3 Just "Stuff ups" or Collective (Economic) Unintelligence?

Economic crises of various types occur reasonably regularly in nations and regions around the world. Marhjnsen [14, p 533] lists fifteen developing nations associated with the 1980's debt crisis. The US financial system with its Savings and Loans crisis was also restructured to accommodate unrepayable debts. As Feenstra and Taylor [15] detail, 1992 saw six European currencies depreciate up to 25 percent. The 1997 Asian exchange rate crises impacted eight nations heavily (peak falls in the baht and rupiah were 50 and 80 percent) with subsequent pursuit of strong external surplus positions (mirrored by high deficit positions in several developed nations such as the USA and Australia). Other notable currency crises include the Brazilian real (-40% in 1998), Russian ruble (-80% in 1999) and Argentine peso (-80% in 2002). De Paoili, Hoggarth et al. [16] list 45 crises between 1970 and 2000 involving default, of which those also involving exchange rate and banking crises numbered 21 with an average length of 10 years and a mean cost per year of 22 percent of GDP. Collective failures in economies are neither rare nor trivial.

"Every crisis is different, of course. Ukraine faced hyperinflation in 1994; Russia desperately needed help when its short-term-debt rollover scheme exploded in the summer of 1998; the Indonesian rupiah plunged in 1997, nearly levelling the corporate economy; that same year, South Korea's 30-year economic miracle ground to a halt when foreign banks suddenly refused to extend new credit. But I must tell you, to IMF officials, all of these crises looked depressingly similar." [17]

At base in a crisis is some unexpected event which sees expectations out of kilter with new realities. Johnson [17] sees the economic solution as "seldom very hard to work out" and focuses on the political and institutional issues which brought about the crisis and may stand in the path of an IMF style solution, a "solution" much debated. Actual transitions from crisis, such as those of the post-Soviet nations in Central and Eastern Europe, demonstrate both variously successful resolutions of crisis and that it is often difficult to effect "a solution". Success is not assured, nor a path determined.

Several distinct types of crisis that can be variously interlinked are recognised in international economics. A convenient reference is the development [15, 16] which combines bank, exchange rate, default and industry crises in an interlinked group, as reproduced below.

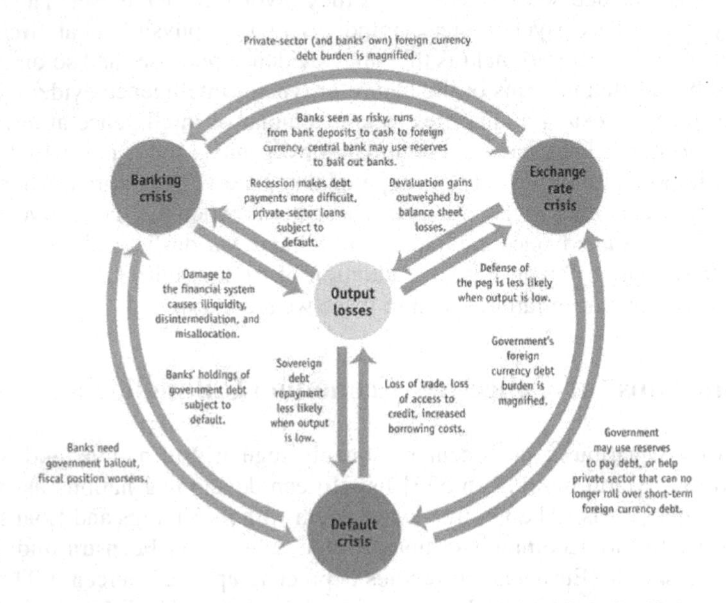

Fig. 1. Vicious Circles in Twin and Triple Crises [15 Fig 22-23]

Single or successively building twin and triple crises are taken as impacting on real output (centred on product production and consumption), with further feedbacks possible from the real economy to the finance, monetary and government sectors.

The elements in the schema are these:

- Banks and a financial system, termed Finance, F say
- A set of currencies variously exchangeable, Monies M
- Governments, G, with their policies and influences "for the collective good"
- Product producers and consumers, as industries I

The whole, which might be termed "FIGM", is a figment of their "informed" imaginations. FIGM relays a pattern[4]. It is an indeterminate complex of four distinct elements variously interrelating.[5] However, not only are these elements interlinked. Each element is itself complex having some internal structure and arrangements that condition conduct and may (or may not) allow for internal resolutions of problems. Any exercises in unintelligence that may lead to crisis can then be seen as potentially

[4] Note that while the pattern is based around four elements there are omitted direct considerations, such as of balance of payments imbalances.

[5] Note that external (to the focal nation) linkages provided by currency, financing, trade and other flows from the rest of the world are not shown. Alternately, the schema may be taken to represent a global situation so the model may be assumed closed externally, but each item (node and flow) is then internally heterogeneous in marked and significant ways.

internal to some element, linked externally in some way to another somehow-like element, and/or interlinked generally with several or all other elements.

Facing the question "How is sense to be made of such confusion - or richness of possibilities?" we make assumptions, including of significant associations and of some sequencing via time or cumulative effect. That is, we build a model. A cadence of interactions or flows is imposed across the complex to produce some applicable order.[6] We then assume that a preferred cadence is relevant and somehow meaningful, or at least sometimes (or at some time) plausible.[7] Much debate centres on the superiority of this cadence or that, both technically and as deemed relevant.

It may surprise but the idea of "crisis" is little considered in "International Economics" texts.[8] It was interesting recently to find how effective introducing "crisis" at the beginning of semester and paralleling it with standard expositions of theory appeared to be in developing student understanding. This mirrors educational experiences in international business and industry analysis where reflections on the tensions between alternate perceptions aided learning. Each of these three areas has a distinctive yet complementary focus and position. Together they can provide richer insights into international and national business situations. Speculatively then, what complementary insights might advance collective intelligence?

4 A Vector Example

Following a short recap of vector basics, triple products are illustratively explored to demonstrate various generated outputs. There appears to be some potential to aid understanding of collective situations in economics and intelligence. The development is preliminary and suggestive so comments are very welcome.

A vector is geometrically a directed line. It can be specified by a length and direction or by a joining of two ordered points. Metaphorically, a base point can be seen as successively projected across space to a destination. Alternately, a directed line is bounded and positioned. Additional assumptions allow "the line" to be "moved across space" if suitable parallels are maintained.

Consider now three vectors **f, g** and **m.** Assume that they cross at some common point O. The vectors can be seen as the distinct projections from some common point O of collocated entities F, G, and M. Two types of vector products can be formed:

- **Scalar** (also known as dot or inner, designated by ".") **product**. One vector is projected on the other with a scalar (or non directed) output. A vector property disappears and a single, unpositioned number remains.
- **Vector** (cross or outer, designated by "x") **product**. One vector is combined with another with a non-coplanar vector output. The output is in a different dimension.

[6] This is essentially the line of development of Cournot and others who build from a sequence of propositions and inferences to suggest possible configurations and inferred conclusions.

[7] For example, currency crises since 1970 have been classed as four generations, with each reflecting a different configuration within Fig 1.

[8] Anecdotally, only one of eight advanced undergraduate texts scanned included explicit discussions of "crisis" or "currency crisis"

With three vectors, compound products are possible. The scalar triple product is of the form **f.gxm** while the vector is of the form **fxgxm**. The former has a single numerical or scalar output (equal to the volume of a parallelepiped built on the three vectors) while the latter has a vector output (in the plane of the two vectors first crossed). Both products synthesise three "distinct elements" into a single or a set of dimensionally different products. The scalar triple product is illustrated in Figure 2.[9] As an illustrative application, this product can be seen as an indicator of credit available at some time C_t Credit is built on projections adopted by financial, governmental and monetary entities. F, G and M together project C.

The point in using vectors and their compound products here is that three separate entities taken "together" (specifically at the same point O) can produce (*via* product operations) a collective outcome which is infeasible if any one was absent or not "co-operative". Not only does a vector project a point across space, three vectors can alternately project the value of a subtended volume and an output vector which is a linear combination of the first crossed pair (despite the generative presence of the third non-coplanar vector). In both cases, the output is dimensionally different to the basis set. The scalar triple product lacks any spatial dimension while the vector triple product lacks any unique dimensional attributes of the procedurally-third vector. These are intriguing outcomes for well-ordered associations of three elements.

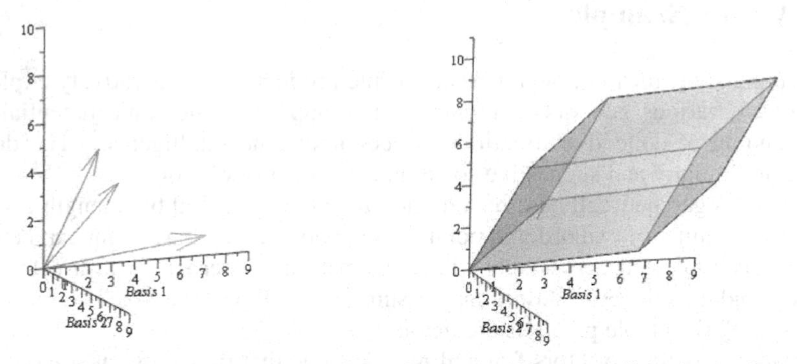

Fig. 2. Scalar triple product with product value projected as a parallelepiped volume

The elements have distinctive contributions and are thus of value and necessary presence but they are not equi-valent. This unequal valency[10] is lost in the collective scalar outcome so some "exogenous" allocation or distribution schema will be needed. In economies this is often couched in custom or culture or concepts such as "a fair go". The area may be one of societal accord or discord.

Speculatively, does a triple product qualify as an expression of intelligence?

– Both definitional requirements of Pfeifer and Scheier appear met, superficially at least, if we accept dimensional change (which is a change in collective quality when the set is combined according to the rules) as a behavioural outcome.

---

[9] Considerations of the vector triple product are left to another place.

[10] Literally, "strength" in some situation of combination. This is interpretable as differing ability to contribute or power of contribution. Note that there are both quantitative and qualitative components of strength.

– The suggested working definition is two-parts met but "successfully" needs elucidation. If, for example, the goals of the vectors or of those cojoining them included having a basis for a volume of a certain size, then the scalar triple product might indeed be adjudged not only "successful" but "intelligent".

Vectors are, of course, assumed to be non-sentient entities yet a case could be built that there is *or may be* intelligence in their triple products. Alternately, perhaps it is not the vectors but the product processes that instills "intelligence"? Is it Nature (of the vectors) or Nurture (via the product environment) or both that yields the outcome adjudged as intelligent – or is the whole situation somehow misspecified? More broadly is there an inference of intelligence by the observer or is the product as an output of human thought an artefact with intelligence embedded. If a triple product is an invention of imagination, did the imagination embed or uncover intelligence? Such questions may lead in many and potentially constructive directions.

Applying such considerations to issues of crisis, a "crisis event" can be seen as a situation when projections by parties based on their expectations are not realized in practice (ie, with the passing of time or influential events). The panel in Fig 3 illustrates an hypothetical credit crisis where expected credit $C_1 = 37$ is reduced in an F "meltdown" to $C_2 = 23$ followed by an increased projection by G to rebuild credit $C_3 = 36.8$, ie to near C_1. Government "projects 80% more" to offset the "75% lesser projection" by finance and the associated 38% drop in Credit available.[11] This illustrates the Keynesian liquidity trap and the governmental response which is ideally short term until the projection **f** of Finance F recovers. That is, credit on issue is maintained by changing the contributions of the elemental parties in offsetting ways.

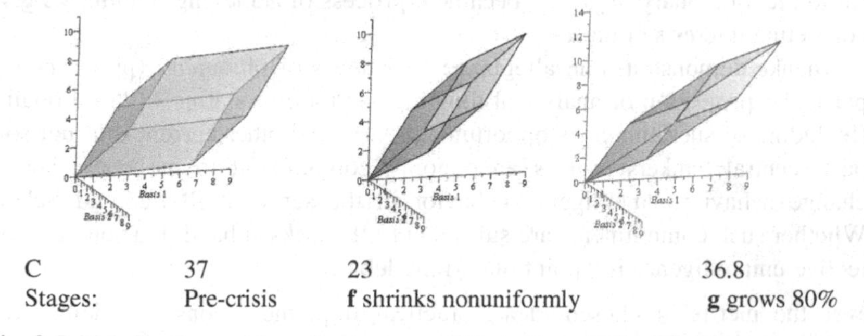

Fig. 3. Initial situation, crisis and response: an illustration of "Credit" differently composed

The schema can also be applied to intelligence if the vectors reflected "assessed intelligence".[12] Each of three parties (or three parts of one party) is set against the three dimensions of some definitional basis, such as those earlier mentioned. For some reason the scalar value of **f** is assessed "downwards", perhaps as a result of some noted unintelligence and/or "meltdown". Within-group compensation is through a greater

[11] Monetary parties M are assumed to remain unchanged for ease of exposition. Note also that government G is assumed to act directly without involving finance F. To the extent that G borrows via F, **g** as projected (but not as obligated) will be reduced and **f** increased.

[12] Presuming some suitable basis, means of measurement and units of measure.

reliance on **g** to maintain the value of the collective product, a reliance that could lead to further crises (Figure 1).

It is an open question whether the collective credit or intelligence C is sustainable or apposite to the situation faced, or whether a fresh start should be made. So far, a "same-as-was-usual" ploy has been the favoured response for both the collective credit and unintelligence problems of the global financial crisis. However, history indicates a marked and lasting reduction in credits and debits along with insolvencies at some stage in any true recovery, a repositioning of ideas and attitudes and new judgments of "creditworthiness " and "intelligence".

5 Conclusion

The approach of Hegel is commonly cast in terms such as "Thesis, Antithesis and Synthesis". Some identified "thing"[13] is associated with a different one resulting in some new thing. Often the preference is to choose "the competing opposite" as the antithesis but the method is wider. One or more "complementary differents" may be used as in the vector example. The process whereby synthesis is achieved is *aufheben*,[14] a term little used or, arguably, understood in English.

Two examples illustrate something of the many usages, and some of the potential problems if "unintelligence" were to be used simply as a competitive antithesis:

- The arguments of Marx and others cast a thesis (Labour, a projection of direct human effort) and antithesis (Kapital, an alternate projection) in competitive opposition, despite their mutual interdependence in production, consumption and societies. The posited struggle between them has coloured much history, and for the revolutionary *aufheben* became a process of achieving dominance given competing interests or theses.
- Bernanke demonstrates an allegiance to a thesis of rationality (preferencing a particular projection of analytical thinking) and a discounting of "irrationality" (including of such things as opportunism). He and other current (but not some past) central bankers profess an ongoing commitment to monetary rules to change behavior, "intelligent" behavior in the sense of Pfeifer and Scheier. Whether such commitments are sufficient to the tasks at hand or a source of collective unintelligence is a point of current debate.

Whatever the merits of chosen ideas, practical implementations can demonstrate much collective unintelligence. Interpretation of influences as competing opposites is but one interpretation, one of arguments in an idiomatically two-vector scalar product tradition. Using triple products could enrich our dialogues and representations.

- Production could become the product of land, labour and capital variously combined "co-operatively" to yield scalar and vector product outcomes.

[13] Such as an entity, attribute, position or projection, for example.

[14] *Aufheben* is a rich word with a range of interpretations. Briefly, it contains the idea of building up while breaking down (as in Schumpter's "creative destruction"), the idea of maintaining some things while changing others (as evident in the "diversity-compliance tradeoff" casting of intelligence by Pfeifer and Scheier), and the idea of opposition and differences to be resolved. The concept deserves renewed attention.

– Central bankers might consider various outcomes of rationality, opportunism and unintelligence (be this in institutional, consumer or institutional stances).
– Using FIGM and like projections, an improved representation of "Credit" or perhaps "Intelligence" and their various interplays may be developed.
– The combinations of Hegel's elements can be cast in a new light.

Triple products can provide rich second stage representations. Whether these are meaningful and usable is the related first stage issue. Both warrant attention.

Efforts to understand collective intelligence are important as are those to avoid or mitigate crises. It is hoped that ideas in this paper illustrate how areas might be advanced not only technically but in ways meaningful to our lives and societies. Clearly much remains to be done. Cooperation and mutual regard in dialogues and developments are needed if the area of collective intelligence is to become potent and enabling. Collective unintelligence and alternative models of association indicate ways to better appreciate crises in economics and how humans with their artefacts "intelligently" position and project themselves to achieve and advance, or otherwise.

References

1. De Bono, E.: Conflicts: a better way to resolve them. In: A Pelican book, vol. viii, p. 207. Harmondsworth, Penguin (1986)
2. Keynes, J.M.: The general theory of employment, interest, and money. Macmillan, London (1936)
3. Krugman, P.R.: Pop internationalism, vol. xiv, p. 221. MIT Press, Cambridge (1996)
4. Pfeifer, R., Scheier, C.: Understanding intelligence, vol. xix, p. 697. MIT Press, Cambridge (1999)
5. Gardner, H.: Frames of Mind: The Theory of Multiple Intelligences. Basic books, New York (1983)
6. Armstrong, T.: Multiple intelligences (2000), `http://www.thomasarmstrong.com/multiple_intelligences.htm`
7. Cournot, A.: Researches into the Mathematical Principles of the Theory of Wealth. Reprints of Economic Classics. Augustus M. Kelley, New York (1960) [1838]
8. Ricardo, D.: On the Principles of Political Economy and Taxation. Dent, London (1969) [1821]
9. Olson, M.: The logic of collective action: public goods and the theory of groups. Harvard University Press, Cambridge (1965)
10. Goleman, D.: Working with emotional intelligence 1998. Bantam, New York (1998)
11. De Bono, E.: Six thinking hats, p. 207. Key Porter Books, Ontario (1985)
12. De Bono, E.: Six Frames for thinking about information. Vermilion - Random House, London (2008)
13. Smith, A.: An Inquiry into the Nature and Causes of the Wealth of Nations. Penguin Books, Ringwood (1986) [1776]
14. Marthinsen, J.E.: Managing in a Global Economy: Demystifying International Macroeconomics. SW Cenage Learning, Mason (2008)
15. Feenstra, R.C., Taylor, A.M.: International Economics. Worth Publishers, New York (2008)
16. De Paoili, B., Hoggarth, G., Saporta, V.: Costs of Sovereign Default. Bank of England Quarterly Bulletin 2006 (Fall 2006)
17. Johnson, S.: The Quiet Coup. Atlantic (2009)

Building Actor Reputation in Web-Based Innovation Networks

Sabine Fliess, Arwed Nadzeika, Marco Wehler, and Jorinde Wormsbecher

FernUniversität in Hagen, Fakultät für Wirtschaftswissenschaften,
Douglas-Stiftungslehrstuhl für Dienstleistungsmanagement, Universitätsstraße 71,
58084 Hagen, Germany
dlm@fernuni-hagen.de

Abstract. In order to be successful in terms of market share, sales, and profit, companies from different industries are detecting the innovative power of the customer network. Handing over more and more elements of the innovation process to the customer is accompanied by a loss of control of the innovating company thus creating quality uncertainty concerning the innovation process. According to New Institutional Economics these uncertainties can be overcome by building actor reputation within the web-based innovation network. Based on a short overview of the different stages of innovation process organization we will show how relevant actor reputation is for innovation networks. We develop an explanatory model of reputation building based on sociological theories of role modelling, interaction and communication and offer first considerations how the model can be tested. We conclude with a summary and an outlook on further research.

Keywords: reputation, networks, web-based innovation, role modelling, evaluation.

1 Introduction

Innovation is one of the crucial elements for economic and entrepreneurial success [1]. To be successful in terms of market share, sales, and profit innovative companies have to combine technological possibilities with market requirements [2].

In the last twenty years, the requirements of innovation management have been changed. Competition is more global, markets are more fragmented, and the increasing individualization of customer needs requires greater product diversity and shorter innovation cycles. Therefore, it is necessary to get product development costs under control and to reduce flop rates.

To reach these aims companies discover the abilities of web-based innovation networks. These networks allow for synchronizing and parallelizing the different phases of the innovation process thus reducing transaction costs and costs of redesigning and testing products. By integrating customers in early phases of the innovation process adaption to customer needs can be ameliorated. Results can be reduced development costs, decreased time-to-market and lowered flop rates.

T.J. Bastiaens, U. Baumöl, and B.J. Krämer (Eds.): On Collective Intelligence, AISC 76, pp. 13–24.
springerlink.com

But handing over parts of the innovation process to users' means at the same time a decrease of control of the innovation process by the innovating company. This can create uncertainties concerning the quality of the innovation process and its outcome. So, web-based innovation networks reduce market uncertainty on one hand and raises innovation process uncertainty on the other.

Uncertainty can be reduced by different means. In this paper we will focus on so called actor reputation. Actor reputation refers to the perceived innovation ability of a member of a web-based innovative network. We assume that identifying actors with a high innovation reputation within an innovation network helps a company to distinguish between successful and non-successful contributions thus accelerating the innovation process, reducing development costs, time-to-market and flop rates.

The aim of the paper is to explain how actor reputation is built during the innovation process thus helping the company to support the process of actor reputation building and to identify actors with a high innovation reputation.

We focus on web-based innovation networks as self-organized networks like virtual communities. Participants of the network are implied to be equal in possibilities and responsibilities. Self organization means that there is no focal leader who organizes a system according to rules or special arrangements. The only rules to be obeyed are those given by the technical system, the platform on which the activities take place. The rules of working together are formed out during the innovation process.

Besides, there are no entry barriers, i.e. everybody who has access to the internet and is willing to reveal his/her name to the network provider can participate. Furthermore, network actors contribute to the network by posting their contributions in forums, chats, on blackboards and the like. There is no third-party organized reputation system [e.g. 3] as it is used by amazon or ebay to evaluate vendors and purchasers. Rather, there are only verbal evaluations.

The paper is organized as follows: First, we will give an overview of the different stages of the innovation process. Based on these insights we will show in the second part of the paper how relevant actor reputation is for innovation networks. In the third section we demonstrate how actor reputation can be built or created during the innovation process within the innovation network. The explanatory model developed in this section is based on sociological theories of role modelling, interaction and communication. The dimensions and determinants of actor reputation will be stressed. In the fourth section we offer first considerations how the explanatory model can be tested. We conclude with a summary and an outlook on further research.

2 Stages of Innovation Process Organization

For a better understanding the innovation process can be analytically sub-divided into different stages [4] [5]. The mapped innovation process referring to Cooper is distinguished by a semi-discrete arrangement of sequences over time (figure 1). Following this model the initial idea has to pass through five gates and correspondingly alternating five stages. In the ideal case the initially poorly conceived idea develops within the innovation process towards a market ready product-launch or a finalized product improvement at the end. The third-generation innovation process is characterized by a fluid mapping, in which the stages as well as the gates are acting like frames depending on the specific requirements of an innovative idea [6].

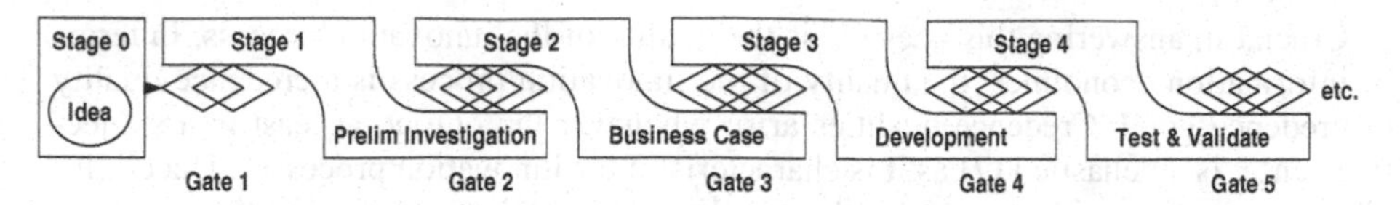

Fig. 1. Third-generation innovation process [6]

Traditionally, innovations and inventions are a responsibility of the R&D department of a company. So, originally all stages of the innovation process have been conducted within the innovating company. As the identification and understanding of customers needs and wants has been identified as a crucial success factor in innovations [7], the idea of integrating the customer into the innovation process came up [8]. Customers can be involved in only one up to all stages of the innovation process [9] [10] [11]. Examples for web-based customer integration into innovation processes are Spreadshirt.com (http://www.spreadshirt.com) where customers design their own t-shirts, ideas4unilever (http://www.ideas4unilever.com/research/ideas4.nsf/index.htm), BMW customer innovation lab (http://www.hyve-special.de/bmw/index1.php), Dell idea storm (http://www.ideastorm.com) or Tchibo Ideas (http://www.tchibo-ideas.de).

Web-based innovation networks permit to integrate different customers into the innovation process at the same time. Based on Cooper's process model a web-based innovation network is considered to consist of actual and potential customers of a company communicating, interacting and collaborating in order to develop new ideas, to create new business models, prototypes, products or services which are to be put successfully on the market.

Using the "wisdom of the crowds" [12] customers create innovative ideas. In an "open innovation" process where not only the process is open to everybody but also the product is accessible to everyone customer-to-customer-networks develop prototypes and final products and create and deliver services to one-another [13] [14] [15].

These networks are a considered to be a cost-efficient way to gain a surplus in effectiveness. Using the advantage of interaction in networks, customers cross-fertilize themselves to new points of view and in the bottom-line to new solutions. This works especially for the creative process of idea generating but also for the evaluation of ideas and first solutions. Networks can bring together different user experiences and impressions as well as practical conceivability of market-needs. For this reason customer integration via web-based networks has the potential to conduce to higher target achievement.

3 Reputation to Reduce Quality Uncertainty in Innovation Processes

The usage of innovation networks is accompanied by handing over more and more phases of the innovation process to the customer. As a consequence the innovative company gives up part of its management responsibility thus "democratizing innovation" [16]. Although the innovative and creative power of the customer is not to be questioned the usage of customer leads to a loss of control of the innovating company thus creating new uncertainties. One of these uncertainties is the question if the loss of control will pay off by increased success rates of new products and services.

Crucial in answering this question is the quality of the innovation process. In terms of information economics, the quality of the innovation process is a credence quality or credence good. Credence qualities arise whenever the output, at least in a subjective sense, is stochastic [17] as it is characteristic for innovation processes. Uncertainties in evaluation arise because of the supplier's uncertainty concerning the representativity of the network members and the quality of their contributions.

According to neo-institutional economics reputation can be regarded as an institution to overcome the quality uncertainty of credence qualities [17]. The theoretically based solution of the uncertainty problem can be supported by the observation that suggestions by well-known developers are rated higher and that contributions by those well-known developers gain more feedback than those of unknown participants. Actor reputation therefore can serve as a selection criterion for high quality contributions thus accelerating the innovation process and reducing selection costs.

4 An Explanatory Model of Actor Reputation Building

4.1 Characteristics of Reputation

Although there are different definitions of reputation, some common body of the theoretical construct can be derived. Reputation refers to a certain salient evaluative characteristic of an entity, e.g. a person, organization or institution, which is estimated to remain stable over time [18] [19] [20]. With respect to innovation networks, reputation refers to the innovation ability of its members. According to the innovation process shown in figure 1 the innovation ability of a certain person refers to the following dimensions:

- the ability to create innovative ideas,
- the ability to develop new business models based on the innovative idea,
- the ability to transfer innovative ideas into prototypes and suitable products and services,
- the ability to outguess market acceptance and market success,
- the ability to evaluate ideas, suggestions, prototypes, products or services in terms of technical feasibility and economic success.

The reputation building process is based on two intertwined and only analytically separable processes: the process of role modeling and the evaluation based process of attribution. Both processes are linked by interaction and communication activities.

As a result of the role modeling process the role an actor plays within the innovation process is defined. Role definition is required to characterize the reputation area. Additionally, roles are used to specify behavioral requirements a network actor has to meet.

Actors differ in their ability of role-playing. Actor's reputation is a measure for his/her ability to fulfill a certain role. The higher an actor's reputation is the more this actor exceeds his/her role requirements.

4.2 Role Modelling

"Roles describe specific forms of behavior associated with given tasks; they develop originally from task requirements" [21]. In an innovation network a role consists of a set of rights, duties, expectations, norms and behavior a person has to fulfill within the different stages of the innovation process. According to different stages of the innovation process different roles can be distinguished.

According to organizational theory "a role consists of one or more recurrent activities out of a total pattern of interdependent activities which in combination produce the organizational output" [21]. Recurrent activities within the innovation process like idea generation, suggestion making, or evaluating, can be combined in a role. In collaboration these roles "produce" the innovation.

Roles are the result of an interaction and communication process called role modeling consisting of an interplay of role-taking and role-making. People take over roles (role-taking) and construct their own roles (role-making) to finally play their particular role (role-playing) [22] [23] [24].

Following the five stages of the innovation process, the following five roles can be identified:

1. Idea generator: generates ideas for new products or services or shows all-day problems,
2. Suggestion maker: makes suggestions for solutions, e.g. suggests how ideas could be transformed into products or how all-day problems could be solved, makes suggestions for business models concerning pricing, distribution or communication, or comes up with solutions, e.g. develops specifications, sketches, prototypes,
3. Forecaster: offers information concerning the acceptance of the solution and potential market success,
4. Evaluator: evaluates ideas, suggestions, models, solutions, prototypes, forecasts,
5. Observer: observes the communication and interaction process without active participation.

The five roles belong to different role types. The first three roles refer to the specific stages of the innovation process; they are active roles to be undertaken by every member of the innovative network. The evaluator role is an active role as well which in contrast to the first three roles can occur at every stage of the innovation process. It can be undertaken by every member of the innovation network. Even the network member whose ideas, suggestions, models, developments or forecasting are to be evaluated can take over this role if s/he is to criticize him/herself. The observer role is a passive role which also can occur at every stage of the innovation process. In reputation building only the first four roles are relevant; the observer role just represents the tacit quantity of the network. Every member of the innovation network can take over one or more roles [25].

In playing their roles the network actors contribute to each of the stages of the innovation process. The success of the innovative network in terms of market convenience and market acceptance is dependent on the role players. The more network members become network actors in taking over one of the first four roles, the more ideas, solutions, prototypes, products and services will be generated. The innovation network's success depends not merely on quantity but rather on quality. The better network actors play their roles the more valuable their contributions are. Actor reputation is the result of an attribution process of other network members evaluating the actor's ability to play his/her role.

4.3 Reputation Building

Reputation is built by an interaction and communication process [26] during which external observers ascribe one or more of the above mentioned characteristics [27] to a network actor. The process of ascribing is based on the communication and interaction activities of the actor. Therefore, the actor's communication and interaction activities have to be public so that they can be observed by different network members. Reputation itself is the result of a collective attribution [28] [29]. While network members observe communication and interaction activities according to the roles of the network actors they evaluate these activities coming to a conclusion concerning the reputation relevant characteristics of the actor's innovation ability. The attribution of the characteristics to the actor forms the basis for the network members' further action, i.e. the attributing network member communicates and interacts on behalf on the attribution. As other network members experience how network actors communicate and interact with respect to the attributed reputation they base their subsequent activities on this observation, either following or negating the attribution.

A social network like an innovation network can be interpreted as a system of different role members [21]. While roles are "standardized patterns of behavior" [21], reputation is linked to a special identifiable person, not to a role.

The reputation building process is based on the different contributions of the network actors as shown in figure 2. Three different types of contribution can be distinguished. According to their relevance for the innovation process and in order of their appearance there are first tier contributions, second tier contributions and third tier contributions. They form the elements of reputation building.

Task oriented first tier contributions. Task oriented contributions are those which refer directly to one of the stages of the innovation process. They consist of

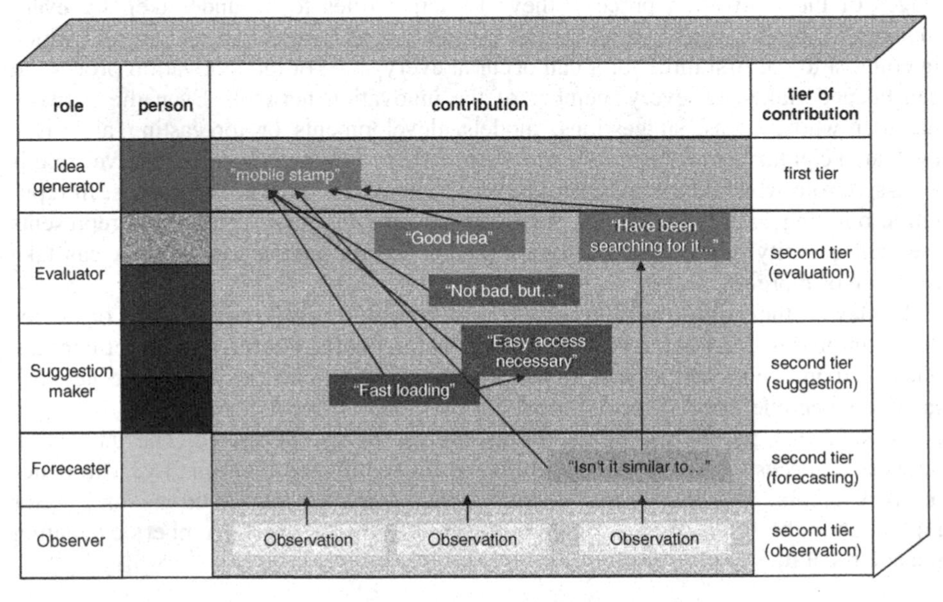

Fig. 2. Interaction of contributions – example

descriptions of problems to be solved, new ideas, suggestions how these ideas could be transferred into solutions, suggestions for business models or parts of business models or information about existing solutions or market acceptance. Task oriented contributions are aimed at the promotion of the innovation process. In web-based innovation networks, task oriented contributions open a new thread of discussion. These contributions can be allocated to the specific stages of the innovation process. Task oriented first tier contributions evocate reactions from other network members.

Task oriented second tier contributions. The reactions to first tier contributions are called task oriented second tier contributions. These contributions can take two forms: evaluations or suggestions.

Evaluations refer to verbal comments of appreciation, affirmation, rejection or disaffirmation. Evaluations help sharpen the roles of the idea generator, the suggestion maker, and the forecaster as the evaluator comments on ideas, remarks and suggestions. Furthermore, evaluations form the basis for actor reputation. The more evaluators come to a positive evaluation of the contributions of the actors the higher the reputation of the role player is. Finally, evaluations cannot only establish reputation of the role players but also promote the innovation process. If evaluative comments motivate the evaluating actor him/herself or one of the other actors to contribute more frequently or to take over an additional role, the innovation process becomes more valuable.

Suggestions are based on a silent evaluation of former contributions which leads to new proposals. As suggestions refine certain elements of former task oriented contributions they incorporate an indirect evaluation of more or less valuable elements of these contributions. Suggestions can be used to bring forward any stage of the innovation process.

Relation oriented contributions. Relation oriented contributions refer to the relations between the network actors. They enhance the social capital of the network. Social capital refers to "features of social organization, such as trust, norms... that can improve the efficiency of society by facilitating coordinated actions" [30]. The higher the social capital, the more network actors trust each others thus encouraging more contributions. An emerging quantity of contributions form a broader basis for establishing actor's reputation [31].

Besides its mere information every communication contains a relationship aspect which defines the relationship between sender and receiver telling the receiver how the message has to be interpreted [32]. Although the transfer of relationship information in an online context is more limited than in an offline context as gestures and intonation are missing at least the kind of words help interpret the assumed relationship on behalf of the sender.

The process of reputation building consists of the interaction of the basic elements.

Reputation building as an interaction and communication process. In an interaction and communication process the three types of contributions show up at different times of the process. Reputation is built by a sequence of task oriented first and second tier contributions. Only positive evaluations and suggestions based on positively evaluated contributions enhance an actor's reputation. So far reputation building has been explained as an direct interaction and communication process of at least two actors.

Reputation can also be regarded as the result of an indirect interaction and communication process [18]. Network actors can decide to observe the reactions of the others before s/he reacts him/herself in order to get a clearer picture of his role in the innovation network. While observing the second tier contributions of different network actors and the reactions towards these s/he gets an impression of the different reputations and thus gains a picture of the hierarchical and relational structure of the network. Thus, by observing the interaction and communication process accompanied by implicit evaluation of the contribution of different actors during the interaction and communication process the reputation information diffuses throughout the innovation network.

Over time first tier contributions and the evaluations and suggestions based on them become obsolete and irrelevant for the innovation process. Thus, a steady flow of contributions of a network actor is necessary to build reputation.

As reputation is area dependent an actor can earn reputation in one or more areas by taking over more than one role. The more areas the reputation is based on the more reputable the actor is. As skills and abilities are diversified it is less likely that a network actor's reputation incorporates all areas. Nonetheless, more than one field of contribution can lead to higher reputation.

If a network actor specifies in one role, e.g. idea generation, suggestion making, or forecasting, reputation is easier to build if s/he becomes involved in more than one innovation process. Timely overlapping processes are to be preferred as not to lose the once earned reputation.

A particular role in reputation building is assigned to relation oriented contributions. Every communication contains a task oriented and a relation oriented part thus defining the actor's position in the network. Relation oriented contributions form the basis for trust between network members thus enabling and facilitating the interaction and communication between network members. As already mentioned above, the quantity and evaluated quality of contributions build up an actor's reputation. So, relation oriented contributions let the reputation building process roll. Figure 3 shows the connection between the different types of contributions.

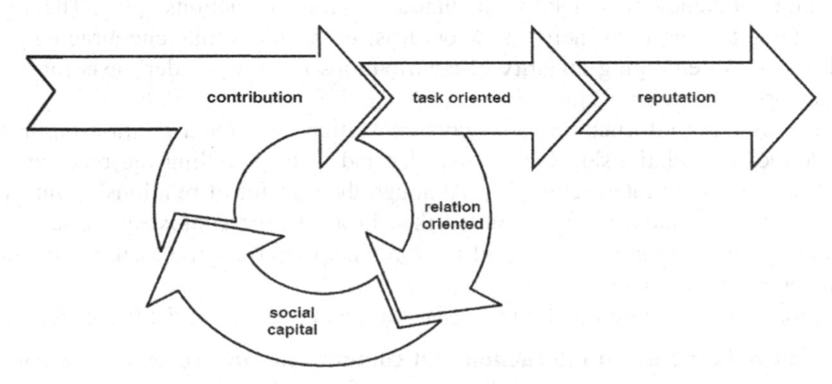

Fig. 3. The wheel of reputation building

5 Suggestions for Empirical Research

Based on these insights, the following hypotheses can be formulated. Reputation is higher

- the more contributions a network actor makes during the innovation process (quantity of contributions during the innovation process)
- the more direct evaluations and suggestions his/her contributions evoke (quantity of evaluations and suggestions during the innovation process)
- the more valuable the contributions are for the innovation process (positively evaluated task oriented contributions)
- the more valuable the contributions are for the relationship building within the network (positively evaluated relation oriented contributions, social capital)
- the more stages of the innovation process the contributions refer to (number of reputation areas)
- the more innovation processes the network actor contributes to (number of innovation processes)
- the more recently the contributions of the reputable actor have been made (discounting of early contributions is necessary).

While most of the variables contributing to an actor's reputation are easy to operationalize (e.g. overall contributions during the innovation process) this is less true for the evaluation of task and relation oriented contributions. A solution to this problem could be Bales' interaction process analysis, a method to study the interaction between people in small groups [33]. According to Bales observed activities of group participants could be classified according to the scheme presented in table 1.

Table 1. Bales' classification system of observable interaction activities

Social-emotional area: positive	Positive reactions	1. shows solidarity, raises other's status, gives help, reward
		2. shows tension release, jokes, laughs, shows satisfaction
		3. agrees, shows passive acceptance, understands, concurs, complies
Task area: neutral	Attempted answers	4. gives suggestion, direction, implying autonomy for other
		5. gives opinion, evaluation, analysis, expresses feeling, wish
		6. gives orientation, information, repeats, clarifies, confirms
	Questions	7. asks for orientation, information, repetition, confirmation
		8. asks for opinion, evaluation, analysis, expression of feeling
		9. asks for suggestion, direction, possible ways of action
Socio-emotional area: positive	Negative reactions	10. disagrees, shows passive rejection, formality, withholds help
		11. shows tension, asks for help, withdraws out of field
		12. shows antagonism, deflates other's status, defends or asserts self

Self-evidently, the different categories have to be adopted to the limited expressions in online communications. Nonetheless, the task oriented expressions can be linked to the task oriented contributions while positive and negative reactions can be partly linked to second tier task oriented contributions and partly to relation oriented contributions, thus lowering or enhancing actor's reputations.

While this scheme is appropriate for the content oriented classification of network actor's contributions it does not refer to the processing of reputation building. To analyze the knowledge flow within an innovation network TeCFlow could be used in combination with interaction process analysis [34].

6 Summary and Outlook

In the previous sections we have shown that the innovating companies experiences quality uncertainty by handing over parts of the innovation process to an innovation network consisting of customers. Actor reputation was considered to be a solution to these uncertainty problems.

Based on sociological theories an explanatory model has been developed consisting of two intertwined processes: role modelling and reputation building. Role modelling refers to the innovation process. Roles are overtaken by the different members of the innovation networks consisting of task oriented contributions to the innovation process. Interaction and communication are used as means for role modelling. Via interaction and communication processes task contributions of different role takers form the actor's reputation. Crucial to reputation building are task oriented second tier contributions consisting of evaluations and suggestions. These activities establish social ties between the network members thus creating social capital within the network. The higher the social capital the more contributions actors make thus moving the "reputation wheel".

Based on this explanatory model hypotheses have been formulated. TeCFlow and interaction process analysis have been suggested as means of empirical testing. This part of the paper has to be further developed. Hypotheses have to be refined. According to our hypotheses reputation is an agglomerated and cumulated construct and could also be named reputation capital. An algorithm to measure reputation capital has to be developed and additional methods have to be considered. Finally, the explanatory model has to be tested empirically.

If the model is going to be supported by empirical results it could form a basis for the development of reputation systems in not only innovation networks.

References

1. Schumpeter, J.A.: The Theory of Economic Development: an Inquiry into Profits, Capital, Interest, and the Business Cycle. Harvard University, Cambridge (1934)
2. Dougherty, D.: Interpretive Barriers to Successful Product Innovation in Large Firms. Organization Science 3, 179–202 (1992)
3. Sorge, C., Zitterbart, M.: A Reputation-Based System for Confidentiality Modeling in Peer-to-Peer Networks. In: Stølen, K., Winsborough, W.H., Martinelli, F., Massacci, F. (eds.) iTrust 2006. LNCS, vol. 3986. Springer, Heidelberg (2006)

4. Crawford, C.M.: New Products Management. Irwin, Burr Ridge, Boston (1994)
5. Hughes, G.D., Chafin, D.C.: Turning New Product Development into a Continuous Learning Process. Journal of Product Innovation Management 13, 89–104 (1996)
6. Cooper, R.G.: Third-Generation New Product Processes. Journal of Product Innovation Management 11, 3–14 (1994)
7. Brentani, U., de Ragot, E.: Developing New Business-to-Business Professional Services: What Factors Impact Performance. Industrial Marketing Management 25, 517–530 (1996)
8. Thomke, S., von Hippel, E.: Customers as Innovators: A New Way to Create Value. Harvard Business Review (April), 5–11 (2002)
9. Kaulio, M.A.: Customer, Consumer and User Involvement in Product Development: A Framework and a Review of Selected Methods. Total Quality Management 9, 141–149 (1998)
10. Magnusson, P.R., Matthing, J., Kristensson, P.: Managing User Involvement in Service Innovation: Experiments with Innovating End Users. Journal of Service Research 6, 111–124 (2003)
11. Reichwald, R., Piller, F.: Open Innovation: Kunden als Partner im Innovationsprozess (Open Innovation: Customers as Partners of the Innovation Process). In: für Erich Zahn, F., Habenicht, W., Foschiani, S., Wäscher, G. (eds.). Peter Lang, Berlin (2005)
12. Surowiecki, J.: The Wisdom of Crowds. Anchor (2005)
13. Chesbrough, H.W.: Open Innovation: Researching a New Paradigm. Oxford University Press, Oxford (2008)
14. Chesbrough, H.W.: Open Innovation: The New Imperative for Creating and Profiting from Technology. Oxford University Press, Oxford (2006)
15. Kozinets, R.V., Hemetsberger, A., Schau, H.J.: The Wisdom of Consumer Crowds: Collective Innovation in the Age of Networked Marketing. Journal of Macromarketing 28, 339–354 (2008)
16. von Hippel, E.: Democratizing Innovation. MIT Press, Cambridge (2005)
17. Darby, M.R., Karni, E.: Free Competition and the Optimal Amount of Fraud. Journal of Law and Economics 17, 67–88 (1973)
18. Büschken, J.: Reputation Networks and Loose Linkages between Reputation and Quality. In: Diskussionsbeiträge der Katholischen Universität Eichstätt, Wirtschaftswissenschaftliche Fakultät Ingolstadt. Katholische Universität Eichstätt, Ingolstadt (2000)
19. Herbig, P., Milewicz, J., Golden, J.: A Model of Reputation Building and Destruction. Journal of Business Research 31, 23–31 (1994)
20. Yu, T., Lester, R.H.: Moving Beyond Firm Boundaries: A Social Network Perspective on Reputation Spillover. Corporate Reputation Review 11, 94–108 (2008)
21. Katz, D., Kahn, R.L.: The Social Psychology of Organizations. John Wiley & Son, New York (1966)
22. Mead, G.H.: Mind, Self and Society. University of Chicago Press, Chicago (1967)
23. Goffman, E.: The Presentation of Self in Everyday Life. Peter Smith Pub., Gloucester (1999)
24. Scott, J., Marshal, G.: Role. In: A Dictionary of Sociology. Oxford University Press, Oxford (2009)
25. Kieser, A., Kubicek, H.: Organisation (Organization), Berlin, New York (2007)
26. Helm, S.: Unternehmensreputation und Stakeholder-Loyalität (Corporate Reputation and Stakeholder Loyalty). DUV, Wiesbaden (2007)
27. Kreps, D.M., Wilson, R.: Reputation and Imperfect Information. Journal of Economic Theory 27, 253–279 (1982)
28. Bromley, D.B.: Reputation, Image, and Impression Management, Chichester (1993)

29. Sjovall, A.M., Talk, A.C.: From Actions to Impressions: Cognitive Attribution Theory and the Formation of Corporate Reputation. Corporate Reputation Review 7, 269–281 (2004)
30. Putnam, R.D.: Making Democracy Work: Civic Traditions in Modern Italy. Princeton University Press, Princeton (1993)
31. McLure-Wasko, M., Faraj, S.: Why Should I Share? Examining Social Capital and Knowledge Contribution in Electronic Networks of Practice. MIS Quarterly 29, 35–57 (2005)
32. Watzlawick, P., Beavin, J.H., Jackson, D.D.: Pragmatics of Human Communication: A Study of Interactional Patterns, Pathologies, and Paradoxes. W. W. Norton & Co., New York (1967)
33. Bales, R.F.: Interaction Process Analysis. A Method for the Study of Small Groups. University of Chicago Press, Chicago (1976)
34. Kidane, Y.H., Gloor, P.A.: Correlating Temporal Communication Patterns of the Eclipse Open Source Community with Performance and Creativity. Computational & Mathematical Organization Theory 13, 17–27 (2007)

An Approach for the Visual Representation of Business Models That Integrate Web-Based Collective Intelligence into Value Creation

Henrik Ickler

University of Hagen, Faculty of Economics, Department of Information Management, Universitaetsstr. 41, 58097 Hagen, Germany
henrik.ickler@fernuni-hagen.de

Abstract. The rise of the so called Web 2.0 changed many classical business models considerably. New or changed business models systematically integrate the customer into the process of value-adding. The customer is not only consumer of products and services. He is rather directly or indirectly part of the production process. In the definitions of Web 2.0 this phenomenon is called collective intelligence. Roughly, in this context collective intelligence can be explained as a general term for user participation and the resulting added value. Examples like the T-Shirt retailer "Threadless" or the open innovation marketplace "InnoCentive" show the potential of these business models. Conventional methods and approaches for the visual representation of business models do not consider this new circumstance. The existing methods and approaches are inadequate, because they do not represent the special features of this kind of collective intelligence. This paper describes what web-based collective intelligence is, to get a common understanding of it and to have a definition for further work. Furthermore, an approach for the visual representation of business models using collective intelligence that represents these special features is presented.

Keywords: collective intelligence, business model, visual representation.

1 Introduction

The rise of the so called Web 2.0 changed many classical business models considerably. New or changed business models systematically integrate the customer into the process of value-adding. The customer is not only consumer of products and services. He is rather directly or indirectly part of the production process. The crucial feature of Web 2.0 is here the high degree of user participation [11] and the results of the participation process. In the definitions of Web 2.0 this phenomenon is called collective intelligence (e. g. [16]). Roughly, in this context collective intelligence can be explained as a general term for user participation and the resulting added value. Examples like the T-Shirt retailer "Threadless" or the open innovation marketplace "InnoCentive" show the potential of these business models. "Crowdsourcing" [10] respectively "interactive value creation" [19], "open innovation" [4], "social commerce" [22] or "wisdom of crowds" [24] are common concepts in these context.

T.J. Bastiaens, U. Baumöl, and B.J. Krämer (Eds.): On Collective Intelligence, AISC 76, pp. 25–35.
springerlink.com

These concepts and the business models using these concepts, show that collective intelligence is already integrated into the production of goods and services. Conventional methods and approaches for the visual representation of business models do not consider this new circumstance. The existing methods and approaches are inadequate, because they do not represent the special features of collective intelligence. The goal of this paper is twofold. The first goal is to describe what web-based collective intelligence is, to get a common understanding of it and to have a definition for further work. The second goal is the development of an approach for the visual representation of business models using collective intelligence that represents these special features and offers an adequate proposal for different target groups using business models. To reach this second goal, an existing approach for the visual representation of business models by Wirtz [30] is enhanced. The underlying research process corresponds to the design science approach of Hevner et al. [9]. According to these design oriented approach a conceptual solution is developed.

After this introduction the term "web-based collective intelligence" is explained in detail. The basis for the explanation is a framework for web-based collective intelligence according to Malone et al. [13]. Following this, the term business model is analyzed and different approaches for the visual representation of business models are regarded. After that, an approach for the visual representation of business models, using web-based collective intelligence in their production process, will be developed based on the approach of Wirtz [30] and the framework of Malone et al. [13]. The paper closes with a short conclusion and an outlook on future research.

2 Web-Based Collective Intelligence

With the rise of the Web 2.0 the World Wide Web (WWW) became more interactive. O'Reilly [16] has already mentioned that the participation of users and collective intelligence are constitutive principles of Web 2.0. Other authors also mention that the changed role of the customer and collective intelligence are main characteristics of the term "Web 2.0" (c.f. [11, 29]). The users create, edit, mix, connect or share content (c.f. [16, 29, 31]). What the authors mean by the term collective intelligence is not explained in detail. The term web-based collective intelligence summarizes the interaction of customers, the resulting collective intelligence and the approaches of user integration based on the web. The exact meaning of this term is explained below.

2.1 Definition of Web-Based Collective Intelligence

A wide range of definitions and interpretations can be found for the term "intelligence". Even single research disciplines like psychology use many different definitions. A consensus about the exact meaning of the term could not be found until now [27]. Generally, there is no doubt about that intelligence is not a real existing phenomenon that can be directly observed. Furthermore, intelligence is a construct deduced from behavior [21]. In general intelligence can be defined as "the ability to learn or understand or to deal with new or trying situations, the skilled use of reason, the ability to apply knowledge to manipulate one's environment or to think abstractly as measured by objective criteria" [14].

In this context intelligence is allocated to one single individual or an intelligent being. The term "collective intelligence" enhances the term "intelligence" explicitly. A collective consists of a "number of individuals or things considered as one group or whole" [15]. Here, intelligence is not the ability of one single individual or thing. It is rather the ability of the whole group of individuals. Smith phrases it as follows: "The notion of collective intelligence (CI) is that a group of human beings can carry out a task as if the group, itself, were a coherent, intelligent organism working with one mind, rather than a collection of independent agents" [23].

In the context of collective intelligence, the construct can be regarded from two different perspectives. It depends on the cooperation of the single individuals within the collective. On the one hand the so called "collective intelligence of the unconnected individuals" exists. On the other hand the so called "collective intelligence of the connected individuals" exists [1]. The collective intelligence of the unconnected individuals describes the intelligence that emerges when the individuals of the collective act independently from each other. In this case the single results of the individuals will be aggregated by an aggregator. The aggregated result is normally better than the simple sum of the single results. Communication or interactions between single individuals do not exist. The single individuals do not have to know that they are part of a collective. The result will be built by the external aggregator. This aggregator is also responsible for the way the single results will be aggregated.

The collective intelligence of the connected individuals describes the collective intelligence that emerges when the individuals of the collective establish a relationship and act with a certain degree of dependency on each other. The underlying mechanism is similar to the mechanism of the so called swarm intelligence of social insects. The action of single insects or individuals is directly affected by the actions of other individuals within the collective. The result is a self-organization process. A famous example of this kind of collective intelligence is the foraging of ants. Based on the communication and interaction via pheromones, the swarm of ants finds the shortest distance between food source and nest [2].

Considering the general definition of intelligence the collective intelligence of the unconnected and connected individuals is defined as the intelligence allocated to a collective or group of individuals. In case of the collective intelligence of the unconnected individuals, intelligence emerges by the combination of an external aggregator. Figure 1 shows this kind of differentiation. The collective intelligence of the connected individuals and the collective intelligence of the unconnected individuals are extremes. A mix of both types is also possible.

The phenomenon of collective intelligence is not a new one. In different situation collectives or groups of individuals (e.g. working groups or families) or animals (e.g. ants or bees) do things that seem to be intelligent. Together they achieve results, a single individual could not achieve. Low cost communication and interaction enabled by the internet now makes it feasible for groups to do many more things than before. In this context web-based collective intelligence finally is defined as the ability of a collective to learn or understand or to deal with new or trying situations, the skilled use of reason, the ability to apply knowledge to manipulate one's environment or to think abstractly as measured by objective criteria, based on the internet and associated technologies. The achieved result of this collective mechanism is a "preferable" result,

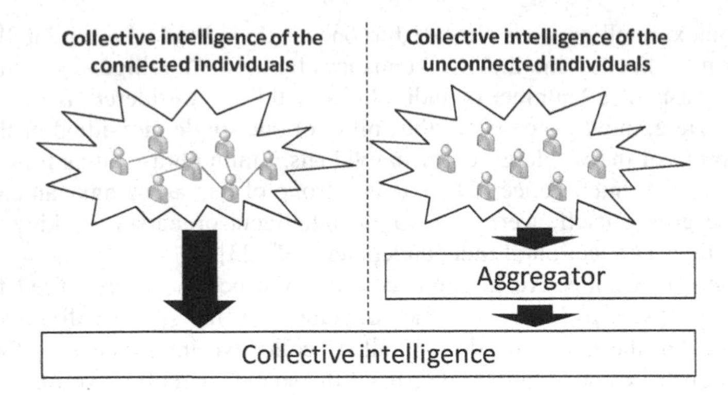

Fig. 1. Two different kinds of web-based collective intelligence

according to the environment. It is usually better than the result of a single individual of the group or a result achieved by conventional methods.

2.2 Framework for Web-Based Collective Intelligence

The definition of web-based collective intelligence implies that it can be found in different variations. Concepts like crowdsourcing already integrate web-based collective intelligence into the production of goods and services. However, a systematic analysis cannot be found in literature. Malone et al. [13] came up with a first approach by splitting web-based collective intelligence into single building blocks called genes. These building blocks can be found in different concepts and approaches that are using web-based collective intelligence. Malone et al. [13] offer with their approach a framework that allows a methodical analysis of web-based collective intelligence. The framework gives a description of the relevant parts of web-based collective intelligence. Knowing these parts makes it possible to take a closer look on business models. Therefore, the framework is used as a basis for the development of an approach for the visual representation of business models that integrate web-based collective intelligence into the value creation process.

To classify the building blocks, Malone et al. [13] use four related questions: Who is performing a task?; Why are they doing it?; What is being accomplished?; How is it being done?

The first two questions ask who undertakes the activity and why the individual takes part in an activity. Within the first question Malone et al. [13] differentiate between "hierarchy" and "crowd". A collective of individuals can be organized hierarchically. Than the single individuals of the collective do not have equal rights. Some individuals do have advanced rights and can undertake activities others cannot. In the crowd all individuals have the same rights and activities can be undertaken by everyone in the collective who wants to do so. An important question is why individuals take part in the activity and what motivates them to participate. As a simplified overview three basic motivators cover the motivations that lead individuals to participate. These motivators are "money", "love" and "glory". For many individuals financial gain is a strong motivator (money). But also intrinsic enjoyment of an activity or the feeling to contribute to a cause larger than oneself can be a motivator (love). Finally, individuals are also motivated when they are recognized by others for their contribution (glory).

The two questions concerning the "what" and "how" are also related to each other. The "what" asks for: What is being done? It can be differentiated between the basic building blocks "create" and "decide". Either individuals generate something new like a piece of text (create) or they evaluate and select alternatives (decide). Furthermore, a differentiation is possible whether individuals do these tasks independently or dependently of each other. For the question about "how" four different building blocks exist. According to the collective intelligence of the unconnected individuals the individuals create something new independently of each other. The result is a "collection" of the single results, by any kind of aggregator. Regarding the collective intelligence of the connected individuals, the individuals also create something in form of a direct "collaboration". A subtype of the building block "collection" is the building block "contest". In contests one or several contributions are designated as the best and receive a special form of recognition like a prize. In a contest the individuals will be motivated by "money" or "glory".

The building block "collaboration" occurs when individuals of a collective work together to create something new and important dependencies exist between their contributions. This building block conforms completely with the description of the collective intelligence of the connected individuals. For the decision tasks also two different building blocks exist. It can be divided into "group decision" and "individual decisions". On the one hand individuals make decisions independently of each other. In this case the decision represents the decision of a single individual. On the other hand decisions exist, which represent the decisions of a whole collective. This group decision can occur in several ways. Important variants are voting, consensus, averaging and prediction markets.

The building blocks of web-based collective intelligence by Malone et al. [13] are summarized in the following Figure 2. They can be found in several combinations by phenomena of web-based collective intelligence.

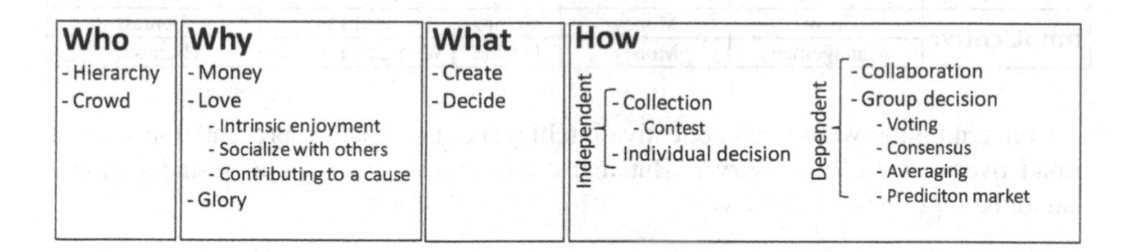

Fig. 2. Building blocks of web-based collective intelligence

To have a better overview about the understanding of web-based collective intelligence and the illustrated building blocks, the relation of the single components of web-based collective intelligence is presented in Figure 3.

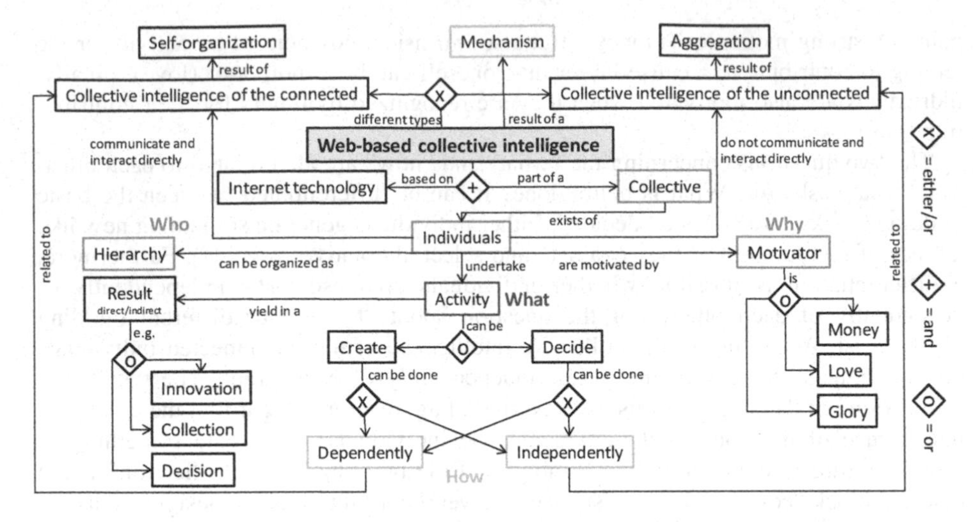

Fig. 3. Web-based collective intelligence

With the help of the building blocks it is possible to represent business models which integrate web-based collective intelligence into the value creation process (from now on known as web-based collective intelligence business models). Because of the differentiation in four related questions, the observer can easily identify who undertakes an activity, why people undertake an activity, what is being done and how the activity is undertaken. The following Table 1 gives an example. It shows the representation of the company "InnoCentive"[1]. The tabular presentation shows that a crowd creates a solution. The reason that leads individuals to participate is a financial award. The basis of the process is a contest. Management is part of a hierarchy within an organization. It gets a financial compensation in form of salary.

Table 1. Building blocks of web-based collective intelligence at the example of InnoCentive

	Who	Why	What		How
InnoCentive	Crowd	Money	Create	Solution	Contest
	Management	Money	Decide	Who gets rewards	Hierarchy

Concerning the web-based collective intelligence, the visual representation gives a broad overview for an observer. But many important aspects of a business model cannot be represented that way.

3 Business Models

The term business model became famous in the context of the so called new economy. In literature and practice no common understanding could be found until now [20]. In

[1] InnoCentive (http://www.innocentive.com) is a so called open innovation marketplace. Organizations can post challenges on the online platform of InnoCentive and offer registered problem solvers significant financial awards for the best solutions.

general, a business model is defined as the representation of how a business works [3] or as the essence of the theory of an organization [12]. Definitions in literature vary strongly and have different focuses. Some authors focus on elements of a business model (e.g. [26]). Other authors focus on the purpose of a business model (e.g. [18]).

3.1 Definition of the Term Business Model

Some definitions of the term business model have achieved broad acceptance in literature. In this paper we will follow the definition of Timmers [26]. According to him a business model is:

- "An architecture for the product, service and information flows, including a description of the various business actors and their roles; and
- a description of the potential benefits for the various business actors; and
- a description of the sources of revenues."

Timmers focuses on the term architecture and the relevant flows within this architecture. His interpretation encloses all actors, their potential benefits and the sources of revenues.

3.2 Visual Representation of Business Models

In the course of time different approaches for the visual representation of business models have been developed. These approaches differ in their goals and purposes and many other characteristics. Especially for the visual representation of e-business models many approaches have been created (e.g. [28, 17, 7]). Moreover, different typologies for business models were mentioned in literature (e.g. [26, 30, 25, 12]). Deelmann and Loos [6] analyzed some of the most popular approaches for the visual representation of business models in detail. One of these approaches is the one of Wirtz [30], which is often mentioned in literature (e.g. [5]). The analysis of Deelmann and Loos [6] shows that the approach of Wirtz can be used for most target groups and for many purposes like simulation or knowledge transformation. Compared to other approaches, this approach shows important details of a business model. Therefore, this approach is adequate for diverse applications and is the basis for next steps.

Wirtz [30] adopted the definition of a business model of Timmers [26] and enhanced this description. He describes a business model as a construct that consists of six single models – market model, procurement model, value creation model, value proposition model, distribution model and capital model. Additionally, he suggests a business model typology that distinguishes between four basic types of business models. These types are "content", "commerce", "context" and "connection". The business model type "content" contains the collection, selection, systematization and appropriation of content. The business model type "commerce" contains all business models that have a focus on business transactions like the initiation or processing of a deal. Business models of the type "context" cover the structuring and classification of information in the internet. Whereas the business model type "connection" contains all business models that focus on the creation of the possibility to share information in networks.

Wirtz [30] uses a self-developed visual notation for the visual representation of business models. This notation is a combination of graphical elements and texts. The single models, all involved actors and their interrelations will be represented graphically. Designations and further descriptions will be represented in text form. Figure 4 gives an example of this kind of visual representation.

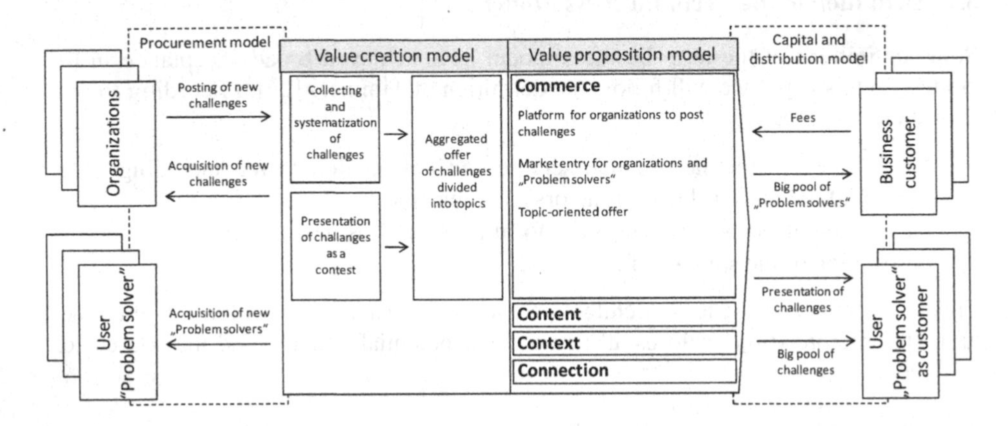

Fig. 4. Visual representation of a business model according to Wirtz

4 Visual Representation of Business Models That Integrate Web-Based Collective Intelligence into Value Creation

The example of the visual representation of a business model according to Wirtz [30] gives a good overview about a business model. Relevant elements and facts, like how value is created or who the customers are, can be seen. Different target groups get the possibility to analyze the business model. Moreover, they can visualize the business model for several purposes [6]. Additionally, homogeneous groups of business models can be built with the presented typology [30]. Overall, Wirtz [30] offers an adequate approach for a visual representation of a business model and a typology. That provides a good basis to analyze and to construct business models. However, this kind of visual representation is developed especially for e-business models. Wirtz and Ullrich [31] showed that this approach can be used to represent Web 2.0 business models. The visual representation focuses on value proposition. The value proposition model will be particularized by the differentiation into the four C's "content", "commerce", "context" and "connection". But the visual representation of web-based collective intelligence business models is not possible in its entirety.

Web-based collective intelligence business models have their distinctive feature in the way they create value. To analyze and construct this kind of business model, the value creation process has to be considered in detail. Value creation does not only take place inside the organization. Rather a group of customers is an external part of value creation. They can perform tasks in different processes, like in management, core or support processes. For this reason a further differentiation of the value creation model makes sense. An obvious differentiation is made into the two types "customers" and "company". The type "company" contains value creation by an organization or by external service providers, whereas the type "customer" contains value creation by a group of customers. Based on this differentiation the visual representation clearly shows who is active in the value creation process.

Furthermore, for a more comprehensive visual representation of the business model it makes sense to integrate the framework of web-based collective intelligence by Malone et al. [13] into the approach. The four related questions are integrated into the different models. The question who is active is presented in the two new types "company" and "customer". The question why people are motivated to participate is presented in the value proposition model. Finally, motivators like "money", "love" or "glory" are a value proposition of an organization to the customers.

The question about what is being accomplished and the question how it is being done are also represented in the value creation model. Therefore, an additional dimension with the name "contribution" is used. This type describes what the contribution of the customer is and how this contribution is.

As a consequence a web-based collective intelligence business model is visual represented with the domains "content", "commerce", "context", "connection", "company", "customer" and "contribution". Figure 5 gives an example of a visual representation of a business model using this seven C's. In comparison with Figure 4 the integration of web-based collective intelligence into a business model becomes clearer. It can be seen in detail where the relevant aspects of web-based collective intelligence in the respective business model are.

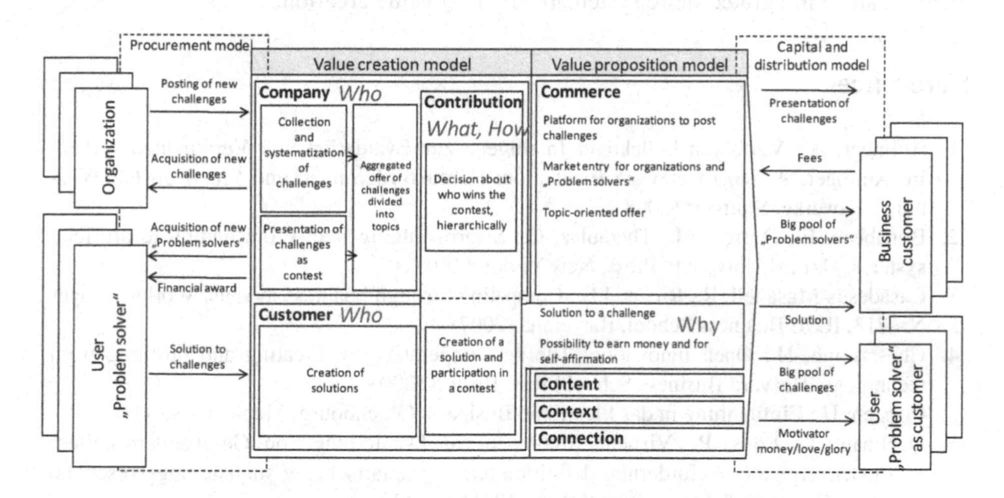

Fig. 5. Visual representation of a business model using seven C's

5 Conclusion and Outlook

The development of the WWW to the Web 2.0 presents itself in strong user participation. The success of business models like "Threadless" or "InnoCentive" shows that it is possible to integrate customers and users into the value creation process in an effective and efficient way. But existing approaches for a visual representation of business models do not represent web-based collective intelligence business models in its entirety. Next to a description of web-based collective intelligence, this paper presents an approach for the visual representation of web-based collective intelligence business models by combining the approach of Wirtz [30] and the framework of Malone et al. [13]. By integrating these two approaches special features of web-based collective intelligence are represented.

As a result, diverse target groups can systematically visualize web-based collective intelligence business models e.g. to analyze them or to develop new business models.

The analysis and development of web-based collective intelligence business models is still in its beginnings. The current situation suggests that this kind of business model will gain further relevance in the future. The presented approach is a first attempt to represent these business models in a visual way. A next step in this research process will be a validation of this attempt to show potential benefits or disadvantages. An approach for the visual representation of a business model gives no answers to questions like: Do web-based collective intelligence business models form new types of business models or do the different forms of web-based collective intelligence make a difference concerning business models? But these are still relevant and unanswered questions.

Need for future research in web-based collective intelligence so exists in many areas. Next to the exploration of the mechanism and principles of web-based collective intelligence business models, it is important to show how web-based collective intelligence can be integrated more systematically into value creation.

References

1. Aulinger, A.: Verfahren kollektiver Intelligenz zur Evaluation von Verbundnetzwerken. In: Aulinger, A. (ed.) Netzwerk-Evaluation: Herausforderungen und Praktiken fuer Verbundnetzwerke, Stuttgart (2008)
2. Bonabeau, E., Dorigo, M., Theraulaz, G.: Swarm intelligence: From natural to artificial systems. Oxford University Press, New York (1999)
3. Casadesus-Masanell, R., Ricart, J.E.: Competing through business models. Working Paper Nr. 713, IESE Business School, Barcelona (2007)
4. Chesbrough, H.: Open Innovation: The new imperative for creating and profiting from technology. Harvard Business School Press, Boston (2003)
5. Corsten, H.: Einfuehrung in das Electronic Business. Oldenbourg, Muenchen (2003)
6. Deelman, T.: Loos. P.: Visuelle Methoden zur Darstellung von Geschaeftsmodellen: Methodenvergleich, Anforderungsdefinition und exemplarischer Visualisierungsvorschlag. Research Group IS & Management, Paper 13, Mainz (2003)
7. Gordijn, J.: E-Business value modeling using the e3-value ontology. In: Curry, W.L. (ed.) Value creation from E-Business models, Butterworth-H., Oxford (2004)
8. Gordijn, J., Osterwalder, A., Pigneuer, Y.: Comparing two business model ontologies for designing E-Business models and value constellations. In: Proceedings of the 18th Bled conference (e-Integration in Action), Maribor (2005)
9. Hevner, A.R., March, S.T., Park, J., Ram, S.: Design science in information systems research. MIS Quarterly 28(1), 75–105 (2004)

10. Howe, J.: The rise of crowdsourcing (2006), http://www.wired.com/wired/archive/14.06/crowds.html
11. Kilian, T., Hass, B.H., Walsh, G.: Grundlagen des Web 2.0. In: Hass, B.H., Kilian, T., Walsh, G. (eds.) Web 2.0: Neue Perspektiven fuer Marketing und Medien. Springer, Berlin (2008)
12. zu Knyphausen-Aufseß, D., Meinhard, Y.: Revisiting strategy: Ein Ansatz zur Systematisierung von Geschaeftsmodellen. In: Bieger, T., Bickhoff, N., Caspers, R., zu Knyphausen-Aufseß, D., Reding, K. (eds.) Zukuenftige Geschaeftsmodelle: Konzepte und Anwendungen in der Netzoekonomie. Springer, Berlin (2002)
13. Malone, T.W., Laubacher, R., Dellarocas, C.: Harnessing Crowds: Mapping the Genome of Collective Intelligence. CCI Working-Paper No. 2009-001, Massachusetts Institute of Technology, Cambridge (2009)
14. Merriam-Webster's Online Dictionary, http://www.merriam-webster.com/dictionary/intelligence
15. Merriam-Webster's Online Dictionary, http://www.merriam-webster.com/dictionary/collective
16. O'Reilly, T.: What is Web 2.0?: Design patterns and business models for the next generation of software (2005), http://www.oreilly.de/artikel/web20.html
17. Osterwalder, A., Pigneur, Y.: An e-Business model ontology for modeling e-Business. In: Proceedings of the 15th Bled eCommerce Conference (eReality: Constructing the eEconomy), Bled (2002)
18. Oesterle, H.: Business engineering: Transition to the networked enterprise. EM – Electronic Markets 6(2), 16 (1996)
19. Reichwald, R., Piller, F.: Interaktive Wertschoepfung: Open Innovation, Individualisierung und neue Formen der Arbeitsteilung. Gabler, Wiesbaden (2009)
20. Rentmeister, J., Klein, S.: Geschaeftsmodelle in der New Economy. Das Wirtschaftsstudium (WISU) 30(3), 354–361 (2001)
21. Roth, E.: Der Intelligenzbegriff. In: Roth, E. (ed.) Intelligenz: Grundlagen und neuere Forschung. Kohlhammer, Stuttgart (1998)
22. Rubel, S.: Trends to watch part II: Social Commerce (2005), http://www.micropersuasion.com/2005/12/2006_trends_to_.html
23. Smith, J.B.: Collective Intelligence in Computer-Based Collaboration. L. Erlbaum Associates, Hillsdale (1994)
24. Surowiecki, J.: The Wisdom Of Crowds: Why the Many are Smarter than the Few and How Collective Wisdom Shapes Business, Economies, Societies and Nations, Doubleday, New York (2004)
25. Tapscott, D., Ticoll, D., Lowy, A.: Digital Capital: Harnessing the Power of Business Webs. Harvard Business School Press, Boston (2001)
26. Timmers, P.: Business models for electronic markets. EM - Electronic Markets 8(2), 3–8 (1998)
27. van der Meer, E.: Intelligenz als Informationsverarbeitung. In: Roth, E. (Hrsg) Intelligenz: Grundlagen und neuere Forschung, Kohlhammer, Stuttgart (1998)
28. Weill, P., Vitale, M.R.: Place to space. Harvard Business School Press, Boston (2001)
29. Wigand, R.T., Benjamin, R.I., Birkland, L.H.: Web 2.0 and beyond: Implications for Electronic Commerce. In: Proceedings of the 10th International Conference on Electronic Commerce, New York (2008)
30. Wirtz, B.W.: Electronic Business. Gabler, Wiesbaden (2001)
31. Wirtz, B.W., Ullrich, S.: Geschaeftsmodelle im Web 2.0: Erscheinungsformen, Ausgestaltung und Erfolgsfaktoren. HMD-Praxis der Wirtschaftsinformatik (261), 20–31 (2008)

Open Science 2.0: How Research and Education Can Benefit from Open Innovation and Web 2.0

Oliver Tacke

Technische Universität Braunschweig, Institut für Organisation und Führung,
Abt-Jerusalem-Str. 4, 38106 Braunschweig, Germany

Abstract. Both, Open Innovation and Web 2.0, are concepts used in commerce in order to support the collaboration of different people and the emergence of new ideas. The approaches can be adapted to science, thus offering new opportunities for research and education. If necessary requirements are satisfied, Open Science 2.0 facilitates e.g. the public development of scientific papers and the conduct of public seminars, both harnessing collective intelligence. This way, it is not only possible to improve the individual outcomes, but also to encourage the exchange between theory and practice.

1 Introduction

Solving problems can be highly inspiring and motivating, and even complicated puzzles can be solved on your own given enough time and ambition. As complexity rises, it becomes more and more difficult, though, and the modern world seems to be full of intricate issues. Finally, you cannot do it all on your own and you depend on additional resources. The traditional approach would be to get in contact with a sage expert to help you. However, several critics emphasize that one person can never possess enough knowledge in order to judge complex situations expediently, and that it may be more appropriate to use the collective wisdom of crowds [11,34].

Taking a closer look at science reveals a similar situation: problems have become more complex and often require a joint effort in order to find a solution. Bozeman and Corley found that some of the most frequent reasons for collaborative research are access to expertise or unavailable equipment, aggregation of knowledge as well as productivity, or simply the pleasure of working with others [3]. In fact, within the last decades, collaboration in science has become more common in various disciplines and has been explored empirically. Hunter and Leahey examined trends in collaboration over a 70 year period, using a random sample of articles that were published in two top sociology journals [15]. They discovered that between 1935 and 1940 only 11% of the observed articles were coauthored, whereas between 2000 and 2005 this was true for almost 50%. This increase in collaborative research is consistent with previous findings from sociology [19], economics [18], political science [8], physics [4] and behavioral ecology [25]. Within the latter field, using Hirsch's h-index for quantifying scientific research output, Pike examined connections between collaboration and

T.J. Bastiaens, U. Baumöl, and B.J. Krämer (Eds.): On Collective Intelligence, AISC 76, pp. 37–48.
springerlink.com

impact [25,13]. Results show that authors with a high scientific impact are those who tend to collaborate widely with others, those who form strong bonds with collaborators, and those who are less likely to be part of a clique.

Besides the benefits for research, Bozeman and Corley also point out that collaborative efforts offer chances for enhancing education and training of students, e.g. lecturers can improve their courses continually by accessing the experience within personal networks, and students can team up for joint learning [3].

The objective of this article is to outline how science could benefit similarly from being open and how research and education can utilize Web 2.0 tools for collaboratively constructed knowledge. Consequently, the approach is called *Open Science 2.0*. Section 2 briefly depicts the theoretical foundation of Open Science, Open Innovation and Web 2.0. Section 3 specifies opportunities and applications for research and education and discusses prerequisites and possible problems. Section 4 provides examples how Web 2.0 tools can be used for joint knowledge construction, and finally, Sect. 5 summarizes the paper's findings and shows some future prospects.

2 Theoretical Foundation

Open Science 2.0 is not to be misunderstood as a whole new concept, but rather, constitutes a combination and concurrence of Open Science, Open Innovation and Web 2.0. The following three subsections will elaborate the definition of *Open Science 2.0* step by step.

2.1 Open Science

In the 19th century, popularization of science was seen as a means by which to provide national education and to overcome outmoded beliefs by rationality. Since then, popularization was put more and more on a level with vulgarization and scientists tended to see their audience in their own kind only but not in the general public; thus a gap developed between universities and society. The vast growth of available information and the ongoing specialization amplified this process [7]. As a consequence, the metaphor of 'research in an ivory tower' arose, proposing isolation from the common people. In addition, the principle of 'publish or perish' pushes scientists to keep their ideas secret until they are published; secrecy and taciturnity have become the primary directives.

Adopting this behaviour wastes much potential for innovation, especially in a society that depends on creation, sharing and usage of knowledge. Hence, new and more intensive structures of communication have to be created to support a wide-ranging transfer of ideas between science and public: science should be opened. This does not only mean sharing prefabricated knowledge with others but also developing a mutual comprehension of problems and to work jointly on subjects relevant to theory and practice [7].

So, on the one hand, Open Science intends to intensify the transfer of knowledge e.g. via public lectures or courses, scientific broadcasts on TV or activities in museums or science centres. It also strongly encourages Open Educational Resources which are understood to comprise content for teaching and learning, software-based tools and services, and licenses that allow for open development and re-use of content, tools and services [9]. But although Open Science supports the concept of Open Access – publishing scientific literature publicly on the internet, free of charge and free of most copyright and licensing restrictions [30] – it does not necessarily comprise it. On the other hand, open scientists give insights to their whole research progress, ranging from idea collection to publication of finished articles. They inform about their scientific activity and reflect their problems openly, inviting others to be part of the problem-solving process. Thus, Open Science explicitly advocates participation of other professionals, students and amateurs, leading to an intertwined construction of knowledge. One may argue that the general public lacks the ability to make a valuable scientific contribution but this point of view ignores the fact that science often explores real-life problems. Thus, scholars can at least benefit from discussing their ideas openly, as they not only ensure relevance in practice, but possibly also see problems from a different perspective. Furthermore, there are even some fields of research that depend on the efforts of amateurs, e.g. astronomy [2].

2.2 Open Innovation

In recent years, the concept of Open Innovation has become known to companies as a new paradigm for developing products and services more efficiently. Instead of relying solely on their own internal research, some firms foster intensive exchange with external sources. The integration of customers or users into the entire development process, in particular, can be significant for the creation of value [27].

Substantially, there are three possible core processes [6]:

Outside-In Process. Valuable ideas cannot only stem from inside the company but also from outside as well. It is intended, therefore, to integrate the distributed knowledge of customers, suppliers, other firms or research institutes throughout the whole process of innovation [5]. In 2007, Cisco Systems used an external innovation competition to find a new business. In the end, more than 2,500 people from 104 countries submitted about 1,200 distinct proposals, leading to the idea for a sensor-enabled smart-electricity grid [20].

Inside-Out Process. Companies use external paths to market by launching spin-offs or start-ups in business areas that do not yet belong to corporate strategy. Furthermore, they can license some of their technology actively to others. Finally, companies can profit by letting spillovers happen on purpose (so-called free revealing), thereby giving up on appropriating future rents from this knowledge through patents or secrecy, which may bring numerous advantages to a corporation [1].

Coupled Process. Merging the Outside-In perspective with the Inside-Out perspective leads to the Coupled-Process. It ranges from jointly finding ideas to commercializing new products and is characterized by give-and-take. IBM, for example, used the Open Source approach for their programming environment called Eclipse. Several companies and individuals collaborated on the complex basic platform, thus saving time to market and increasing the rate of standards adoption. Still, they competed on individual products or services that can be offered in addition [23].

Comparing Open Innovation to Open Science reveals that they build on the same fundamental idea: making the boundaries between you and your environment more permeable can lead to a better outcome, since solutions that nobody could possibly predict might emerge. Although both concepts are applied to different domains, they show various analogies (e.g. solving complex problems, intense urgency for innovation, etc.), and therefore, are assumed to follow very similar principles. Hence, findings about Open Innovation will be carefully adopted for Open Science and vice versa while still taking account of possible differences.

2.3 Web 2.0

The term *Web 2.0* is not defined consistently but it is commonly associated with web-based applications enabling the socialization of content. Those tools facilitate communication and the collaborative creation and usage of information spread on the Internet, harnessing openness and collective intelligence. They are easy to use and thus remove the distinction between producer and consumer – in Web 2.0, by using a standard web browser, virtually everyone can participate in the construction of knowledge [24,33]. Typical generic classes of applications are blogs, wikis, online community websites and media-sharing platforms but often functions and characteristics are blended, making it difficult to distinguish between them.

Using Web 2.0 for implementing Open Science and Open Innovation seems to be a natural approach, as they share the same properties such as openness and participation of a wide range of people. In conclusion, the term *Open Science 2.0* does not mean a new version of Open Science but the application of Web 2.0 services and principles of Open Innovation to the domain of research and education.

3 Application and Prerequisites

In research and education alike, Open Science 2.0 can be used throughout the entire process of problem-solving, which focuses ideas. It can be separated roughly into three main phases with specific tasks [32]: Phase number one comprises the generation of ideas that describe your problems, finding new ideas and proposing your own ideas to others. The next phase covers the exploration and evaluation of ideas. The aim is to meld them into a plan that can probably solve the problem. The final phase is concerned with the implementation of particular ideas, their transmission to the recipient and eventually the examination of acceptance and feedback.

There is a large variety of ways how Web 2.0 applications can be used during these phases to utilize collective intelligence: reflections about one's own work, reviews about conferences visited or new ideas can be presented in blogs and discussed publicly in the commentary section. Thus, it is possible to gather supplementary input from different people, reducing the risk of ignoring important facts. Collaborative elaboration can take place in wikis, which offer a flexible and easy-to-use working environment. Members of social networks can be asked to undertake peer reviewing in order to assure quality and to provide important feedback for further improvement. The circle is complete when commenting on articles starts, e.g. in blogs.

For utilizing collective intelligence in research and education, certain requirements have to be satisfied. In general, these have been investigated by Tapscott and Williams [31] and Surowiecki [34] and can be combined for Open Science 2.0:

Being Open. Openness is the fundamental requirement in order to benefit from using Web 2.0 applications in research and education. Certainly, some circumstances can be obstructive such as legal or monetary issues. In science, the motto 'publish or perish' has led to a self-serving system in which people are reluctant to share their thoughts publicly before they have been publicized – someone might steal their 'intellectual property', come up with a paper first and harvest all the fame and glory. On the one hand, this point of view neglects the fact that only few things that can be read in scientific literature actually originate from the author, as they have only been possible by building upon the work of others [17]. On the other hand, if you openly spread your ideas on the Internet, you can prove easily that you were the first one who had the idea. In consequence, the 'pirate' would not only have to fear legal measures but also punishment by the scientific community.

Some individuals might also be afraid of publicly admitting or making mistakes and hence losing prestige in their community. This applies to students participating actively in lectures or projects, to teachers having particular problems in class and to researchers as well. This fear can be hearkened back to the idea that making mistakes is always bad, although they could also be understood as chances to learn. Open Science in general, therefore, requires an attitude towards life that admits that one may be wrong, others may be right and that one perhaps could approximate the truth jointly, as Popper described his philosophy of critical rationalism [26].

One more reason for openness being a fundamental requirement can be derived from the properties of a community that is based on mutuality. If you want others to share their ideas with you and offer help, you have to act alike. Taking an exclusively Outside-In Perspective, e.g. browsing blogs or borrowing the homework from others, may result in some additional input, but will not allow the full potential to unfold. It is crucial to contribute something to social networks if you want to benefit from them.

Finally, being open also means being open-minded. Bringing many stakeholders together can result in the emergence of conflicting ideas. They should be discussed without prejudice and without a precise goal in mind beforehand. Giving up some control and abandoning a devised plan can be more beneficial than sticking to it. This means a higher level of uncertainty but something completely new and unexpected can emerge.

Endorsing Diversity. A wide range of different perspectives can help to find better solutions to a problem, and so by making research and education public, you can gain access to many thoughts, ideas and opinions that you would miss otherwise. The more people that offer support, the better the outcome may become, so it is important to maintain a large network and activate it in due time. Web 2.0 applications seem to be particularly suitable for this intention because they even allow people with little technical skills to contribute their opinions. Those would possibly be missing otherwise. To ensure diversity, it is also necessary for people to specialize and to draw on local knowledge. This means that they have to act in a decentralized way so as to prevent the emergence of groupthink [10]: if you always discuss problems with the same fellow students, it is likely that you will not consider all possible alternatives to a solution, as the desire for unanimity exceeds the desire for quality decisions.

Another aspect related to the independence of opinions is hierarchy. Within the present system of research and education, there is a vast gap between professors, scientific assistants and students. Knowledge that is presented from a 'higher authority' is seldom questioned or contributed to, even though Humboldt demanded an exchange of ideas between lecturers and listeners 200 years ago [14]. To support innovative ideas, it is crucial to create a non-hierarchical environment and it can be necessary to restrain one's own personal ambitions. For example, lecturers should not think of themselves as superior to the students but as a *primus inter pares*, as part of a network. This way they can establish an atmosphere where ideas can emerge unhindered because the supposedly inferiors do not remain silent.

Merging Opinions. Finally, all the ideas have to be aggregated and individual thoughts have to be melded into a collective decision. This process is particularly difficult and adequate methods for this process yet have to be found. Nonetheless, Web 2.0 applications offer several platforms for letting many people cooperate.

In addition to these requirements for harnessing collective intelligence, according to Martin [21], persons solving problems in groups have to possess a sensitivity for networks. They have to be able to grasp the interdependencies occurring cognitively and emotionally. Web 2.0 applications, such as online community platforms for social networking, can support this sensitivity by unearthing the structure of links between persons and groups.

Most of the requirements described above can be satisfied by acting upon the neuron metaphor [22], which interprets humans as neurons of a brain obeying certain rules. They do not hold back their knowledge but 'fire' new impulses as soon as a certain action potential has been exceeded. That does not imply spreading any thought crossing one's mind, but rather, conceptualizing a proper statement and then feeding it to the non-hierarchical neural network without being afraid of embarrassing oneself. In addition, humans seen as neurons should react quickly and should also try continuously to connect to other neurons. This can be achieved by providing much information about oneself and offering motivating projects, thus giving others chances to find common ground for cooperation. Finally, neurons accept uncertainty and bear it.

4 Examples

Subsequently, I will present two examples illustrating how Web 2.0 applications are already used in research and education. The first example reflects my personal experience of writing scientific papers according to the methods mentioned above, and can be connected with action research [12]. The second example describes the experiences of Spannagel and Schimpf [29] and Wiley [35] with public seminars or open teaching, giving students the opportunity to learn in a context-sensitive environment.

4.1 Writing Scientific Papers

Scientific literature is a well-established medium for spreading information and can be considered a resource for learning. However, opening the entire process of scientific construction of knowledge can be even more valuable – being able to participate is not only more motivating for others, but also enables them to learn firsthand, instead of just reading finished compositions.

In order to gather ideas for new projects, I read several scientific and non-scientific blogs and discussed the articles with the authors. I also read the latest papers to keep up with progress, but I observed that discussions on the Internet are by far more up to date and offer a broader perspective. In presenting my premature ideas, I use a public wiki at Wikiversity, which serves as my lab notebook[1];it also contains the first idea collection to this very paper. Usually, I begin with a brief description of the initial point and then I sketch out the contemplated course of the study, followed by a short list of sources that might be useful. The bottom of the page contains a discussion area where anyone can comment or add suggestions.

Of course, in most cases, it does not suffice to make your ideas public; you have to encourage people actively to team up with you. Firstly, I advertise my sketches in several social networks of which I am a member. Secondly, I feed information

[1] My public lab notebook can be found at http://de.wikiversity.org/wiki/Benutzer:O.tacke.

into Twitter[2] [28], requesting comments. If the worst comes to the worst, there is no reply at all; in comparison to the ordinary process of writing scientific papers, this situation is neither better nor worse. However, at best, those who read my message forward it to others, thereby spreading my information even further. Usually that is already enough to stir up enough interest to gain some valuable input.

In this particular case, in the early phase of idea generation I received three hints from two persons, both professors of pedagogy. In the wiki, they recommended several resources related to the subject, some of which I actually used in the end. Besides the page connected with this very article, there is also a section where I openly collect ideas for projects or papers. One of them deals with publishing students' papers and received feedback from three different people, among them one of the professors mentioned above, one student of business information systems and one scientific assistant which I did not know before. They made four suggestions which I will have to further evaluate when I eventually tackle the project. There are also some pages with ideas that have not received feedback so far, but I have not asked for either.

While progressing, I 'tweet' brief status reports, findings or difficulties. This way, on the one hand, others can learn about results before the paper is finished and published. On the other hand, I can receive further input or assistance in return continually. For example, when I publicly stated that writing papers in English was hard for a non-native-speaker, I received an offer for help[3] within a few minutes. Apparently, this approach mimics the Coupled Process from Open Innovation in commerce.

After preliminarily finishing this article, I posted it to a special social networking platform[4], which fosters the open exchange and feedback between people interested in science - not necessarily scientists. I also announced the availability via Twitter, asking for an open peer review. This way, I intended to assure quality once more before submitting the paper, and in fact I received valuable comments from two people, among them one student and one researcher in the field of media education. The suggestions from the latter were particularly useful as they pointed out weaknesses which I would not have noticed otherwise.[5]

This approach could also be applied to term papers or theses. In fact, few days ago, a student of mine announced that he was going to use a wiki to elaborate a term paper completely in public. One might argue that it will not be clear whether he really wrote it himself but that is quite normal and ignores possible advantages: firstly, my student can benefit from the same effects that I described above. Furthermore, I cannot only examine the final result of his efforts but I

---

[2] Twitter is a popular microblogging service facilitating a swift spreading of information, sending messages means 'to tweet', see http://twitter.com.

[3] See http://twitter.com/otacke/status/6017200536 and http://twitter.com/mons7/status/6017385208

[4] See http://wissenschaftler20.mixxt.de

[5] Additionally, I would be glad to receive public feedback from as many readers or participants of the conference as possible in order to learn.

can also monitor the development progress, giving me more information about problems about which I should possibly take care: I can learn about particular difficulties in which students engage. Last but not least, the 'scientific literacy' of the public can be fostered this way.

4.2 Public Courses

Independently, Spannagel[6] and Wiley[7] opened some of their academic courses that had been held traditionally so far. Students were encouraged to post their homework on their personal, publicly accessible blogs and to discuss their ideas in a wiki open to the public. Both lecturers then invited others to join via blogs, Twitter and personal invitations. As a result, several people joined in and debated with the students.

In the particular case of Spannagel, his seminar's subject was a special method of teaching (LdL)[8] which was intended to be adopted for didactics of computer science. The students had to find out the theoretical foundations of the method for themselves and discussed their findings in an open wiki. Spannagel reported about the seminar in his blog and and via twitter and did not foresee what would happen: external individuals joined the discussions about LdL, among them even its inventor, Jean-Pol Martin. This experience was highly motivating for the students and they suddenly discussed issues with Martin's pupils in nineth grade. Additionally, a trainee teacher learned about the special method from the students' wiki and tried it in one of his courses. Finally, he and Spannagel's students worked together: while he reported about his experience from practice, they developed teaching modules he could use. The results are still available[9] and can be used freely.

The added richness of broader perspectives from the outside led to higher motivation for the students and the informal participants of the course could hugely benefit as well. All this innovation could not have happened in that way if the teaching had not been opened to public discussion via Web 2.0 applications, allowing for the use of diverse knowledge from outside.

The approaches of Spannagel and Wiley reflect ideas of Personal Learning Environments (PLEs) that are a quite new approach to using technologies for learning. Whereas dominant Learning Management Systems, such as Moodle or Sakai, focus on the provision of prefabricated information and tools within a course context, PLEs are concerned with enabling a wide range of individual contexts to be adapted to the users' needs. Furthermore, they soften the distinction between the capabilities of learners and teachers by allowing any user to both consume and publish content – just as Spannagel's students did in their wiki. Additionally, PLEs are concerned with sharing resources, not protecting

[6] Christian Spannagel is a professor of mathematics, computer science and teacher training at the University of Education in Heidelberg.

[7] David Wiley is an associate professor of instructional psychology and technology at Brigham Young University.

[8] LdL is the abbreviation of 'Lernen durch Lehren' (learning by teaching).

[9] See http://de.wikiversity.org/wiki/Kurs:Fachdidaktik_Informatik

them [36,16]. These characteristics resemble to the attributes of Web 2.0 services described in section 2.3.

5 Conclusion

In this paper, possible benefits of applying Open Innovation and Web 2.0 to research and education were outlined. Open Science 2.0 embraces collaborative construction of knowledge and can lead to more diverse input for solving problems and also to more motivation for the participants. However, in order to yield the best results, several requirements have to be satisfied. The stakeholders need to be open-minded, willing and able to share their ideas unreservedly without fear of embarrassment or 'intellectual theft'. In addition, diversity has to be endorsed, and so, a wide range of different perspectives has to be appreciated, regardless of the hierarchical position of the contributors. Furthermore, there have to be adequate mechanisms for aggregating the different ideas and merging them into a solution.

Web 2.0 applications facilitate communication and the collaborative creation and usage of information. The Web 2.0 community is just as much based on openness and mutuality and constitutes an ideal environment for the approach of Open Science if the participants can develop a sensitivity for networks. Acting upon the neuron metaphor by Martin is a possible way to gain this competency.

One of the difficulties with Open Science 2.0 is that it contradicts the prevailing notion of research and education in some respects. It is sometimes even regarded as second-class science. Although the two examples presented in this paper show that it is already applied occasionally and that it leads to good results, more convincing research is necessary and critical issues have to be clarified, such as dealing with plagiarism or coping with the dominant 'publish or perish' system mentioned in section 3. Another problem could be seen in the amount of time which has to be spent for networking with others, sharing ideas etc. in order to benefit from collaboration, but then again, there is no such thing as free lunch.

As future work, it is planned to integrate open microblogs in educational events in order to research how knowledge of students and informal participants from the outside can be combined and complex problems can be solved using collaborative intelligence.

Acknowledgements

This paper was improved by conversations with several people. Particular thanks to Jean-Pol Martin, Christian Spannagel and Björn Hobus for many interesting discussions about research and education in general. I would also like to thank Mandy Schiefner and Alexander Perl for peer-reviewing and Christine Charcholla and Monika König for proof-reading the paper.

References

1. Alexy, O.: Free Revealing: How Firms Can Profit From Being Open. Gabler, Wiesbaden (2009)
2. Anderson, C.: The Long Tail: Why the Future of Business is Selling Less of More. Hyperion, New York (2008)
3. Bozeman, B., Corley, E.: Scientists' collaboration strategies: implications for scientific and technical human capital. Reserarch Policy 33(4), 599–616 (2004)
4. Braun, A., Gómez, I., Méndez, A., Schubert, A.: International co-authorship patterns in physics and its subfields, 1981-1985. Scientometrics 24(2), 181–200 (1992)
5. Chesbrough, H.W.: Open Innovation. Harvard Business School Press, Boston (2003)
6. Enkel, E., Gassmann, O.: Neue Ideenquellen erschließen: die Chancen von Open Innovation. Marketing Review St. Gallen 26(2), 6–11 (2009)
7. Faulstich, P.: Öffentliche Wissenschaft. In: Faulstich, P. (ed.) Öffentliche Wissenschaft – Neue Perspektiven der Vermittlung in der wissenschaftlichen Weiterbildung, transcript, Bielefeld, pp. 11–52 (2006)
8. Fisher, B.S., Cobane, C.T., Vander Ven, T.M., Cullen, F.T.: How Many Authors Does It Take to Publish an Article? Trends and Patterns in Political Science. PS 31(4), 847–856 (1998)
9. Geser, G. (ed.): OLCOS Roadmap 2012 (2007), `http://olcos.org/cms/upload/docs/olcos_roadmap.pdf`
10. Janis, I.L.: Victims of groupthink: a psychological study of foreign-policy decisions and fiascoes. Houghton, Mifflin (1972)
11. von Hayek, F.: Die Anmaßung von Wissen: neue Freiburger Studien. Mohr, Tübingen (1996)
12. Hearn, G.: Action research and new media: concepts, methods, and cases. Hampton, Cresskill (2009)
13. Hirsch, J.E.: An index to quantify an individual's scientific research output. Proceedings of the National Academy of Sciences of the USA 102(46), 16569–16572 (2005)
14. Humboldt, W.v.: Über die innere und äussere Organisation der höheren wissenschaftlichen Anstalten in Berlin. In: Gebhardt, B. (ed.) Wilhelm von Humboldts gesammelte Schriften, vol. 10. de Gruyter, Berlin (1968)
15. Hunter, L., Leahey, E.: Collaborative Research in Sociology: Trends and Contributing Factors. The American sociologist 39(4), 290–306 (2008)
16. Kerres, M.: Potenziale von Web 2.0 nutzen (2006), `http://mediendidaktik.uni-duisburg-essen.de/system/files/web20-a.pdf`
17. Luhmann, N.: Die Wissenschaft der Gesellschaft. Suhrkamp, Frankfurt am Main (2002)
18. Maske, K.L., Durden, G.C., Gaynor, P.E.: Determinants of scholarly productivity among male and female economists. Economic Inquiry 41(4), 555–564 (2003)
19. Moody, J.: The Structure of a Social Science Collaboration Network: Disciplinary Cohesion from 1963 to 1999. American Sociological Review 69(2), 213–238 (2004)
20. Jouret, G.: Inside Cisco's Search for the Next Big Idea. Harvard Business Review 87(9), 43–45 (2009)
21. Martin, J.-P.: Wissen gemeinsam konstruieren: weltweit. Lernen und Lehren – Zeitschrift für Schule und Innovation in Baden-Württemberg 33(1), 29 (2007)
22. Martin, J.-P.: Wie verhalten sich Neuronen? (2009), `http://www.adz-netzwerk.de/wiki/index.php?title=Benutzer:Jeanpol/Folie_3&oldid=497`

23. Milinkovich, M.: Eclipse Open Innovation Networks (2007), http://www.eclipse.org/org/foundation/membersminutes/20070920MembersMeeting/07.09.12-Eclipse-Open-Innovation.pdf
24. O'Reilly, T.: What is Web 2.0? (2005), http://www.oreilly.de/artikel/web20.html
25. Pike, T.W.: Collaboration networks and scientific impact among behavioral ecologists. Behavioral Ecology 21(2), 431–435 (2010); Harvard Business Review 87(9), 43–45 (2009)
26. Popper, K.R.: Die offene Gesellschaft und ihre Feinde. Falsche Propheten: Hegel, Marx und die Folgen, vol. 2. Mohr, Tübingen (2003)
27. Reichwald, R., Piller, F.: Interaktive Wertschöpfung: Open Innovation, Individualisierung und neue Formen der Arbeitsteilung. Gabler, Wiesbaden (2009)
28. Simon, N., Bernhardt, N.: Twitter: mit 140 Zeichen zum Web 2.0. Open Source Press, München (2008)
29. Spannagel, C., Schimpf, F.: Öffentliche Seminare im Web 2.0. In: Schwill, A., Apostolopoulos, N. (eds.) Lernen im Digitalen Zeitalter – Workshop-Band: Dokumentation der Pre-Conference zur DeLFI 2009, pp. 13–20. Logos, Berlin (2009)
30. Suber, P.: Open Access Overview (2007), http://www.earlham.edu/~peters/fos/overview.htm
31. Tapscott, D., Williams, A.D.: Wikinomics: How Mass Collaboration Changes Everything. Porfolio, NewYork (2008)
32. Thom, N.: Innovationsmanagement. Schweizerische Volksbank, Bern (1992)
33. Vossen, G., Hagemann, S.: Unleashing Web 2.0: From Concepts to Creativity. Morgan Kaufmann, Amsterdam (2007)
34. Surowiecki, J.: Die Weisheit der Vielen. Goldmann, München (2007)
35. Wiley, D.: Open Teaching Multiplies the Benefit but Not the Effort (2009), http://chronicle.com/blogPost/David-Wiley-Open-Teaching/7271
36. Wilson, S., Liber, O., Johnson, M., Beauvoir, P., Sharples, P., Milligan, C.: Personal Learning Environments: Challenging the dominant design of educational systems. In: Tomadaki, E., Scott, P. (eds.) Innovative Approaches for Learning and Knowledge Sharing, EC-TEL 2006 Workshops Proceedings, pp. 173–182. Milton Keynes, Open University Press (2006)

A Social Network System for Analyzing Publication Activities of Researchers

Alireza Abbasi and Jörn Altmann

Technology Management, Economics and Policy Program
Department of Industrial Engineering
College of Engineering, Seoul National University,
599 Gwanak-Ro, Gwanak-Gu, Seoul 151-744, South-Korea
abbasi@snu.ac.kr, jorn.altmann@acm.org

Abstract. Social networks play an increasingly important role in knowledge management, information retrieval, and collaboration. In order to leverage the full potential of social networks, social networks need to be supported through technical systems. Within this paper, we introduce such a technical system. It is called AcaSoNet. It is a system for identifying and managing social networks of researchers. In particular, AcaSoNet employs a combination of techniques to extract co-author relationships between researchers and to detect groups of persons with similar interest. Past systems have used either search engines to extract information about social networks from the Web (Web mining) or have required people's effort to enter their relationships to others into the system (as being done by most social network services). AcaSoNet, instead, uses a combination of these two types, thereby achieving data reliability and scalability. It extracts and collects data of researchers from the Web but allows researchers to modify the data. In the current version, our system can identify the social network based on publication lists and evaluate the publication activities of users within an academic community.

Keywords: Social network systems, academic community, co-author relationship, publication analysis, productivity analysis, knowledge sharing, knowledge transfer, Web mining, performance analysis, and social network analysis.

1 Introduction

The first social networking Website, Classmates, appeared in 1995. But only in 2003, the Web has become a space for the majority of users to socialize. That year has seen the rapid emergence of a new breed of Web sites, collectively referred to as Social Networking Systems (SNS) [14]. Friendster was the pioneer, while Facebook, Orkut, Yahoo, Google and Microsoft started similar services. They brought structure into the process of personal information sharing and online socialization. However, all these systems are not used in professional environments. It is not possible to use these

T.J. Bastiaens, U. Baumöl, and B.J. Krämer (Eds.): On Collective Intelligence, AISC 76, pp. 49–61.
springerlink.com

systems for collaboration activities, which require the sharing of information in a more sophisticated manner.

Currently, there is no social-network-based system that provides services for the academic community. The social network used for such a system could be based on the collaboration activities (e.g., co-authorship) of researchers. The services could comprise the search for researchers, the recommendations of scholarly articles (based on the authority of the information creator), the automatic building of research communities, the posting of papers to a group of researchers, and the productivity assessment of a researcher and academic community. In particular, many of these services would require the analysis of the social network of co-authorships. In literature, only a few prototypal systems for finding researchers with similar interest have been developed so far [4] [11] [13] [14] [15].

To make the first step towards a sophisticated social-network-based system for researchers and academic communities, we develop AcaSoNet. AcaSoNet provides basic services that enable researchers to analyze their research performance with respect to the number of publications and the collaboration activities.

The challenge that social network systems for academic communities face is the lack of data about scholars, which can be accessed easily. Therefore, the development of the social network system also requires the introduction of a method for collecting data about scholars. Our method is a combination of Web mining techniques for extracting scholarly articles from the Web and manual verification of the mined data by the scholar. The system allows the scholar (user) to modify, delete, and update any data mined. This method guarantees that we only work with user-verified data. Based on this data, AcaSoNet identifies the relationships of researchers within the academic community and evaluates the researcher's performance with respect to the number of publications and the number of co-authorships. In particular, AcaSoNet generates a performance report about the researcher's publication and collaboration activities.

The remainder of this paper is organized as follows: Section 2 gives a brief introduction of social networks and social network analysis. Section 3 discusses existing social network systems in science, their advantages, and their shortcomings. AcaSoNet is introduced in Section 4. This section includes a description of the challenges, design decisions, and the user interface. Section 5 concludes the paper with a brief evaluation.

2 Social Network Analysis and Its Application in Research

Social networks operate on many levels, from families up to the level of nations, and play a critical role in determining the way problems are solved, organizations are run, and the degree to which individuals succeed in achieving their goals. The benefits of the analysis of those social networks are that it can help people to share professional knowledge in a simple way and to evaluate the performance of individuals, groups, or the entire network. For example, with respect to the performance evaluation, the social network of a researcher can be considered an indication of his collaboration activity within a research community [1] [2].

For the analysis, each social network is represented as a graph, which is constructed of nodes and links. Nodes, which denote individuals, organizations, or information, are linked, if one or more specific types of relationships (e.g. financial exchange, friends, trade, and Web links) exist between them. For example, if nodes represent people, a link between two nodes means that two people know each other in some way.

To understand social networks and the impact of certain nodes within the network, the location of nodes (actors) in the network can be evaluated. A social network can be investigated in order to detect collision of interest [4], calculate a person's trustworthiness [7] [8], and to predict referrals between people. For example, if a user explicitly declares friendship to some people, the user will most probably introduce those friends to each other [13]. Furthermore, by determining the node in the center of a social network, social networks can be used to identify experts and authorities on a specific topic [13] [14] [17].

In general, social network analysis (SNA) comprises the measuring of relationships and flows between nodes of a social network. It provides both a visual and a mathematical analysis of human-influenced relationships. Therefore, a social environment can be expressed as a pattern in relationships among interacting actors [19].

In traditional scientometrics literature, social networks in science are investigated by collecting data manually through interviews or questionnaires. Sometimes, the data of those social networks is also obtained through the analysis of co-authored papers and co-citations in scientific publications [5] [10]. A few scientific works have also been performed using data obtained from the Web. The results suggest that real-world networks of academic research communities are closely reflected on the Web [9].

3 Social Network Systems in Science

Beside the approaches for analyzing social networks of researchers, a few social network systems (SNS) have also been developed. These systems not only collect and analyze data but also require active participation of researchers. A few of those systems even provide some basic, Web-based support for collaboration activities. Table 1 gives an overview about those systems [4] [11] [13] [14] [15]. The shortcomings and challenges of those systems face are:

Restricted Use of Collected Social Network Information. A lot of information collected within a SNS is controlled by SNS owners. For example, although FOAF profiles can be created by users, they can only be posted on the SNS Web site, on which it has been created, and linked to other users of the same SNS. Sharing of profiles across domain boundaries is not allowed in many cases. Sometimes, the export of FOAF profiles in machine readable format is not possible, even though one of the purposes of SNS is information sharing [6] [16].

Table 1. Existing social network systems for the academic community

Project (Reference)	Focus of the Project	Methodology Applied for Data Collection
Referral Web (Kautz et al. 1997 [11])	Collects information about the social network of researchers.	After giving the names of researchers as input, names of other researchers are extracted from the Web by searching for the co-occurrence of researcher names on Web pages, using queries against existing search engines.
(Miki et al. 2005 [15])	Measures the value of researchers and their works in the community.	The SNS conducts a bibliographical citation analysis of data found on the Web.
Flink (Mika 2005 [14])	Extracts, aggregates, and visualizes online social networks of communities of researchers.	This SNS employs a co-occurrence analysis of researchers by using queries against existing search engines. Information sources are Web pages, emails, publication archives, and FOAF profiles.
(Aleman-Meza et al. 2006 [4])	Detects conflicts of interests among potential reviewers and authors of scientific papers.	It integrates actors and relationships from two different social network databases, namely from a FOAF social network and the co-authorship network of the DBLP bibliography.
POLYPHONET (Matsuo et al. 2006 [13])	Web-based system for identifying researchers with similar interest at a conference.	It uses search engines, especially Google, to find information about co-occurrences of researchers.

Difficulty in Finding Up-to-Date Social Network Information. In order to find social network information on the Web, most of the above-mentioned SNS use Web mining techniques. However, the likelihood of finding valuable information is very low due to the sheer size of the Web. Since many Web sites are created every day, it becomes difficult for crawlers to find up-to-date information.

Large Number of Queries Against Search Engines. Using search engines for detecting the relationships between users requires a huge amount of queries. In the above-mentioned SNS, the detection of relationships of 500 people would require more than 124000 queries [13]. This clearly shows that the method is not scalable.

4 Concept of AcaSoNet

AcaSoNet is a Web-based, social network system for the academic community. It extracts social network information about researchers from the Web and provides services that leverage the social network information. Currently, AcaSoNet facilitates the execution of performance measures, the management of citations, and the management of publication lists. In the long run, it is planned to extend AcaSoNet by collaboration services. In particular, it is planned to support search for researchers and scholarly articles, as well as the dissemination of research-related information within a social network of a researcher.

4.1 Design Principles

For obtaining social network information, two different approaches are used by existing social network systems: First, users of a SNS state their relations to others by filling out questionnaires or FOAF profiles, linking them to other members of the SNS. Second, social network systems automatically detect relationships through mining of databases. They can mine various sources of information such as e-mail archives, schedules, and the Web [3] [15] [18]. They also analyze keyword co-occurrences within these databases [11] [13] [14].

Since both types of existing data collection methods have their limitations (as explained in the previous section), we propose a method, which is a combination of these two types. In a first step, users provide their exact names and, then, the system queries search engines to find the user's publications on the Web. In a subsequent step, the user can modify and validate the obtained data to make sure that the data stored in the system is correct. Finally, AcaSoNet builds the social network of the user and integrates it into the existing social network of all researchers.

This method produces a precise data set and avoids inconsistent data (i.e. no ambiguity of data). Consequently, the data about users is reliable, while keeping the overhead for the user low. Our approach also reduces the number of queries to a search engine significantly, since it only requires queries for finding information about a person but does not require search engine queries for extracting relationships between users. For example, for extracting co-author relationships, our system does not query search engines. Instead, it asks users to validate the co-author relationships found on the user's publication list. For example, in order to identify co-authors of user A, the list of papers of user A is analyzed. If a co-author B has been found, the system searches for user B within AcaSoNet. If it finds a match, the system will check whether a paper of this co-author has the same title. If the check is positive, B has been identified as the correct co-author of A, and a link between researchers A and B is set. If user B has no account with AcaSoNet, a new account is created. Because of this method, our approach is scalable with the number of users registered with AcaSoNet. In addition to this automatic matching support, each user has the chance to check and modify those links anytime.

Another design principle is that researchers have full control about their social network information. All social network information and the user profile are machine readable and can be exported to other systems anytime. It allows users to use the information for other purposes then the one mentioned here. This means that the effort of researchers put into building up their profile is not wasted, reducing the cost for signing up with AcaSoNet.

4.2 Architecture

The design of the AcaSoNet architecture follows a service-oriented architecture (SOA), using Web services. Since Web services comprise a concept for achieving machine-to-machine interaction over a network and can be executed on remote systems, it allows AcaSoNet to use external services (e.g., a Web search service, the

MyTiesTo service), add easily new services, and to make its own services available for external use. Currently, the only services integrated are the *Portal*, the *MyTiesTo Service* and the *Publication Activity Analyzer* (Fig. 1).

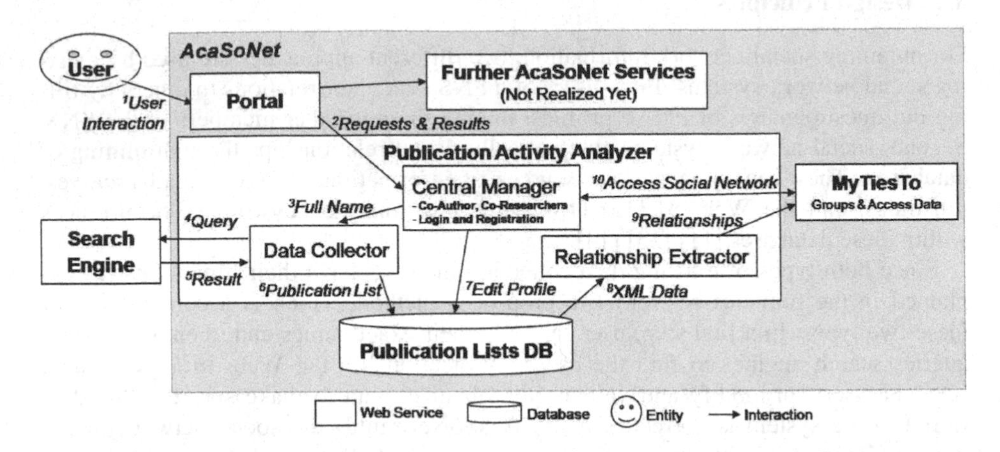

Fig. 1. The AcaSoNet Web service architecture and the interactions between its components

The *Publication Activity Analyzer*, as one service of AcaSoNet, comprises three components (Fig. 1): *Central Manager*, *Data Collector*, and *Relationship Extractor*. The *Central Manager* provides the basic interfaces to external services and coordinates the workflow of the Publication Activity Analyzer. It interacts with the user via the portal (step 1 of Fig. 1). In particular, it handles user requests such as requests for viewing co-author information or for performing analyses of the users' performance (step 2). Currently, the Central Manager can calculate different research productivity indexes (e.g., h-Index, g-Index, RP-Index [2]), the number of collaborations, the collaboration index for researchers (RC-Index) [1], as well as the Co-Author Collaboration Value (which is defined as the total number of collaborations between researchers multiplied by the co-author's productivity index [1]).

In order to update the user profile, the Data Collector (Fig. 1) takes the user's full name as input (step 3 in Fig. 1) and generates a query (step 4) that is sent to a *Search Engine* service. After obtaining the result from the search engine (step 5), the Data Collector parses the results to build the user's publication list, which is stored in the *Publication List DB* database (step 6). Since users keep control about their data, users can modify the publication list via the Central Manager (step 7 in Fig. 1) and the Portal (step 1). Users can also download their profile data anytime. The Relationship Extractor takes the publication list in XML format from the database (step 8) and extracts the co-author relationships of the user. These relationships are stored using the MyTiesTo service (step 9). The Central Manager accesses these relationships (i.e. the social network information) for analyzing the user's performance and presenting it to the user (step 10).

MyTiesTo, another service of AcaSoNet, manages relationships of social networks. It provides an open architecture that allows storing any kind of social network information. In particular, it stores information about entities of the social network as well as the social relationships between entities. Other services can also use MyTiesTo for storing social network information. These social networks can be independent to the social network that the Publication Activity Planner entered. In addition to this, other research-related services (e.g., collaboration support services, chat applications, file exchange services) can make use of the stored social network information for their own services.

4.3 Database Schema

The type of data needed by the Publication Activity Analyzer can be divided into two groups: The first group comprises relationship information (e.g., information about group memberships and communities), which is provided by the MyTiesTo service. The second group comprises information about publications, which are stored in the Publication List DB database. Fig. 2 shows the relationships of the three major tables of the database schema: *Publications* table, *Users* table, and *Publication_User* table.

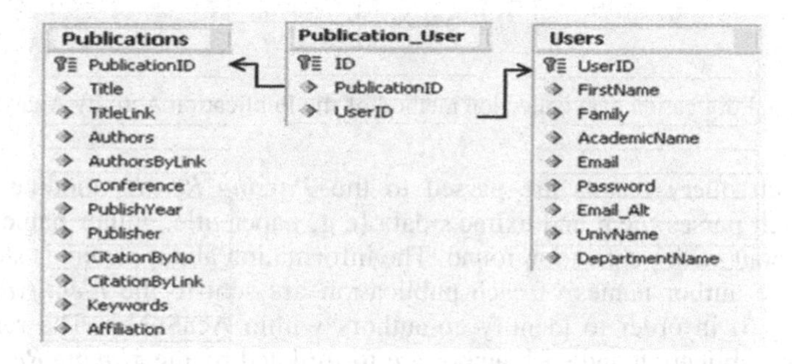

Fig. 2. Database schema of the Publication List DB database

The Publications table stores the information about publications (e.g., journal articles) such as the paper title, the names of authors, the name of the conference or journal, at which the paper had been accepted, and the publication year. The Users table stores the information about a researcher using AcaSoNet. This information includes the user's affiliation, contact information, and name. It is mainly used for identifying the AcaSoNet user. The Publication_User table stores relationships between entries in the two previously mentioned tables. It relates a unique *PublicationID* of a Publications table record to a unique *UserID* of a *Users* table record. Because of this design, information about authors of a paper, papers of an author, or about co-authors can be extracted through simple queries.

4.4 Publication Data Collection Method

Publication data is extracted from the Web by using Google Scholar, a dedicated search engine for scholars. Google Scholar considers a large variety of publication types (e.g., proceedings of national and international conferences, journals, books, and presentations), which can be found in freely accessible databases. Thus, it finds more publications and citations per researcher than other services [12].

In order to query Google Scholar (step 3 in Fig. 3), the Publication Activity Analyzer (i.e. the *Making Query* module) generates a default query (step 2), which is defined as "paper author:name-family". The term *paper* specifies that only papers (not any other type of data like citations or books) should be retrieved. The term *name* represents the author's given name and the term *family* represents the author's surname. Author names and Author family names are retrieved from the Publication List DB database (step 1).

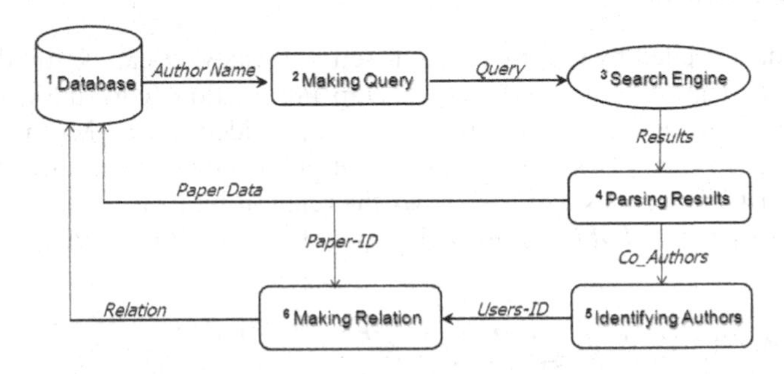

Fig. 3. Publication data collection method of the Publication Activity Analyzer

The search query results are passed to the *Parsing Results* module (step 4 in Fig. 3), which parses them and extracts data (e.g. paper title, author names, publication year) about each publication found. The information about papers is stored in the database. The author names of each publication are sent to the *Identifying Authors* module (step 5) in order to identify co-authors within AcaSoNet. The relationships between the publication and each author are formulated by the *Making Relation* module (step 6) and, then, stored using the MyTiesTo service.

4.5 Further AcaSoNet Services

Currently, AcaSoNet provides only two services (i.e. the Publication Activity Analyzer and MyTiesTo). However, it is envisioned to extend AcaSoNet by offering the following services: *Scholars Search Engine*, *Group Management*, and *Co-Researcher Finder*. The Scholars Search Engine service searches for universities, research institutes, scholars, and experts in a specific research area. The Group Management service allows creating communities of users to which information can easily be distributed. The Co-Researcher Finder service automatically creates groups of researchers based on their research areas. It notifies those groups, if research-relevant data is added or a new researcher joins.

4.6 User Interface

AcaSoNet allows any user (i.e. even a user, who is not registered with AcaSoNet) to access publicly available data. However, if users decide to register, they can, after logging into the system with their login name and password, access more information. Fig. 4 shows the Web site that users see after logging in. The Web site shows a user's publication list, which can be edited. Furthermore, at the top-left side of the Web site, index values (e.g., h-Index, g-Index) and statistics about the user's publications are presented. At the top-right side, a link to the user profile and a link to the list of all co-authors are given. Besides, the user can search for publications to be added to the current list of his publications.

Pub-No	Cit-SUM	Cit-AVG	h-index	g-index	Rp'-index	Rp-index
52	1423	27.36	17	37	10	10.99

Verified Publication List

Welcome Peter Brusilovsky
(Peter Brusilovsky, P. Brusilovsky)
Edit Profile Information

Add Paper

Search for more Publications

List of Co-Authors

		Index	Title	Authors	Year	Citation	Conference
Edit	delete	1	Adaptive hypermedia.	P. Brusilovsky,	2001	316	User Modeling and User Adapted Interaction, Ten Year Anniversary Issue (Alfred Kobsa, ed.) 11 (1/2), 87-110 [PDF].
Edit	delete	2	ELM-ART: An adaptive versatile system for Web-based instruction	G Weber,P. Brusilovsky,	2001	193	InternationalJournal of Artificial Intelligence in Education 12(4), 351-394
Edit	delete	3	Course sequencing techniques for large-scale web-based education	P. Brusilovsky,J. Vassileva,	2003	96	International Journal of Continuing EngineeringEducation and Lifelong Learning 13 (1-2)
Edit	delete	4	From adaptive hypermedia to THE adaptive Web	P Brusilovsky,M. T. Maybury,	2002	81	Communications of the ACM 45 (5), Special Issue on the Adaptive Web, 31-33. [ACM DL]
Edit	delete	5	Adaptive Systems for Web-Based Education	P. Brusilovsky,N. Henze,E. Millan,	2002	79	International Journal of Artificial Intelligence

Fig. 4. Screenshot of the *Verified Publication List* Web page, which the user sees after logging in

Fig. 5 depicts the Web page for searching the Web for publications to be included in the user's publication list. The default query, which is shown initially, can be modified. Any new publication can be added to the list by simply selecting it and pressing the "Add Selected to Verified List" button. The user can also mark query results (i.e. publications) for indicating that those results should not be shown in the future. This is useful in order to avoid seeing publications, which do not belong to the user, every time a new query is submitted. It reduces the sorting overhead for the user. For this, the user has simply to click on the "Never Show in Future" button.

Welcome Peter Brusilovsky
(Peter Brusilovsky, P. Brusilovsky)

Query: paper author:Peter-Brusilovsky From 2000 - To: Search

New Publications Found

		Index	Select	Title	Authors	Year	Citation	Conference
Edit	delete	1	☐	A framework for performance evaluation of user modeling servers for Web	V Zadorozhny, M Yudelson, P Brusilovsky	2008	3	Web Intelligence and Agent
Edit	delete	2	☐	A two-level adaptive visualization for information access to open-corpus	J Ahn, R Farzan, P Brusilovsky	2006	7	Social Navigation and Community-Based
Edit	delete	3	☐	A user modeling server for contemporary adaptive hypermedia: An evaluation of	M Yudelson, P Brusilovsky, V Zadorozhny	2007	9	LECTURE NOTES IN COMPUTER
Edit	delete	4	☐	Adaptation in the Context of Explanatory Visualization	TD Loboda, P Brusilovsky	2008	0	Proceedings of the 3rd European conference on

Never Show in Future

Add Selected to Verified List

Fig. 5. Screenshot of the New Publication Found Web page

By clicking on the name of a user, the Web site with the user profile comes up. It consists of the contact information, the password, the current affiliation, and the previous affiliations. This information is editable, if the profile belongs to the logged in user.

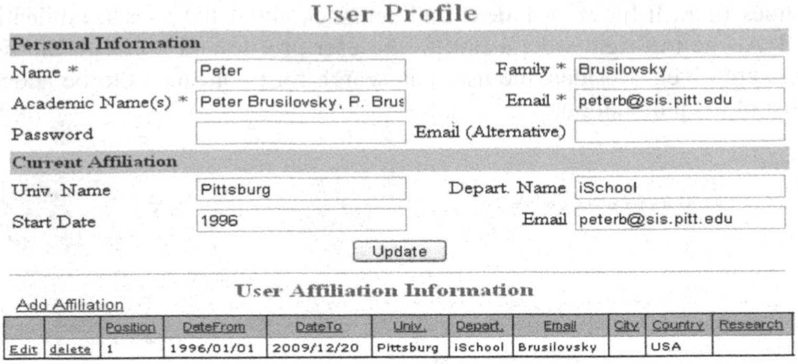

User Profile

Personal Information

Name * Peter Family * Brusilovsky
Academic Name(s) * Peter Brusilovsky, P. Brus Email * peterb@sis.pitt.edu
Password Email (Alternative)

Current Affiliation

Univ. Name Pittsburg Depart. Name iSchool
Start Date 1996 Email peterb@sis.pitt.edu

Update

User Affiliation Information

Add Affiliation

		Position	DateFrom	DateTo	Univ.	Depart.	Email	City	Country	Research
Edit	delete	1	1996/01/01	2009/12/20	Pittsburg	iSchool	Brusilovsky		USA	

Fig. 6. Screenshot of the *User Profile* Web page

By clicking on the affiliation of a user (e.g., Pittsburg in Fig. 6), the research community of the user comes up. Fig. 7 shows the list of researchers that have the same affiliation as the user. The researchers of this research community are either already registered with AcaSoNet (light background) or have received an automatically created account. Note, AcaSoNet creates an account, if a user has not registered yet but is a co-author of a registered user. Unregistered users are shown with a dark gray background.

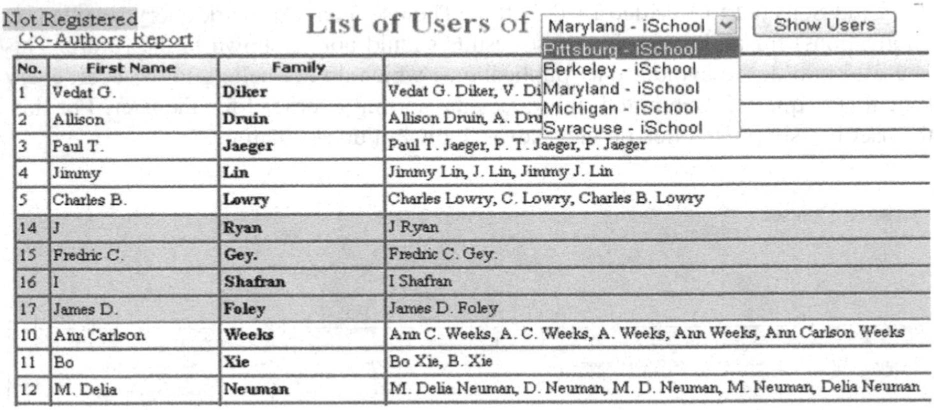

Not Registered
Co-Authors Report

List of Users of Maryland - iSchool | Show Users

Pittsburg - iSchool
Berkeley - iSchool
Maryland - iSchool
Michigan - iSchool
Syracuse - iSchool

No.	First Name	Family	
1	Vedat G.	Diker	Vedat G. Diker, V. Di
2	Allison	Druin	Allison Druin, A. Dru
3	Paul T.	Jaeger	Paul T. Jaeger, P. T. Jaeger, P. Jaeger
4	Jimmy	Lin	Jimmy Lin, J. Lin, Jimmy J. Lin
5	Charles B.	Lowry	Charles Lowry, C. Lowry, Charles B. Lowry
14	J	Ryan	J Ryan
15	Fredric C.	Gey.	Fredric C. Gey.
16	I	Shafran	I Shafran
17	James D.	Foley	James D. Foley
10	Ann Carlson	Weeks	Ann C. Weeks, A. C. Weeks, A. Weeks, Ann Weeks, Ann Carlson Weeks
11	Bo	Xie	Bo Xie, B. Xie
12	M. Delia	Neuman	M. Delia Neuman, D. Neuman, M. D. Neuman, M. Neuman, Delia Neuman

Fig. 7. Screenshot of the *Users with the Same Affiliation* Web page

Finally, the user can get some summary statistics about his/her collaboration activities with other researchers, if the user follows the link "List of Co-Authors" shown on the screenshot of Fig.4. The corresponding Web site not only presents the summary statistics but also the list of co-authors, the total number of publications with each co-author, and some collaboration index values (Fig. 8). By clicking on the total number of publications with a co-author ("No. co-Papers" in Fig. 8), the joined publications are listed in detail.

Welcome Hassan Karimi
(Hassan Karimi, H.A. Karimi,...)

List of Co-Authors & No. of Publications

CoAuthor-No	Collaboration-No	RC-index	rkta-index
17	26	4	4.84

No.	First Name	Family	Academic Name	No. co-Papers	RKCa-index	CCV-index
1	Xiong	Liu	X. Liu, Xiong Liu,	3	1.8	5.4
2	A.	Hammad	A. Hammad	3	1.99	5.97
3	Shuo	Liu	S. Liu, Shuo Liu,	3	0	0
4	L. W.	Yang	L. W. Yang	2	1.8	3.6
5	I.	Bahar	I. Bahar	2	1.8	3.6
6	J.	Peng	J. Peng	2	0	0
7	C. J.	Jursa	C. J. Jursa	1	1.8	1.8
8	M.	Holliman	M. Holliman	1	1.8	1.8

Fig. 8. Screenshot of the *List of Co-Authors* Web page

In the future, we plan to extend the analysis of the collaboration activities of a user and community. For example, we plan to visualize the collaboration network, its structure, and the position of the user within. Currently, however, AcaSoNet can only provide an initial report about the co-authorship network of a user. As shown in Fig. 9, the report gives only an overview about the co-authorship network of a research community in matrix format.

List of All Authors and their Co-Authors & No. of Publications from Maryland - iSchool Show Report

	V. Diker	A. Druin	P. Jaeger	J. Lin	C. Lowry	A. MacLeod	D. Oard	J. Preece	D. Soergel	A. Weeks	B. Xie	M. Neuman	M. White	# of Publications	# of Co-Authors	# of Collaborations
V. Diker	5	-	-	-	-	-	-	-	-	-	-	-	-	5	0	0
A. Druin	-	31	-	-	-	-	-	3	-	4	-	-	-	38	2	7
P. Jaeger	-	-	14	-	-	-	-	-	-	-	-	-	-	14	0	0
J. Lin	-	-	-	37	-	-	2	-	1	-	-	-	-	40	2	3
C. Lowry	-	-	-	-	9	-	-	-	-	-	-	-	-	9	0	0
A. MacLeod	-	-	-	-	-	2	-	-	-	-	-	-	-	2	0	0
D. Oard	-	-	-	2	-	-	61	-	8	-	-	-	-	71	2	10
J. Preece	-	3	-	-	-	-	-	23	-	-	-	-	-	26	1	3
D. Soergel	-	-	-	1	-	-	8	-	21	-	-	-	-	30	2	9
A. Weeks	-	4	-	-	-	-	-	-	-	3	-	-	-	7	1	4
B. Xie	-	-	-	-	-	-	-	-	-	-	15	-	-	15	0	0
M. Neuman	-	-	-	-	-	-	-	-	-	-	-	8	-	8	0	0
M. White	-	-	-	-	-	-	-	-	-	-	-	-	8	8	0	0

Fig. 9. Report about the collaboration (i.e. co-authorship) network of a research community

In detail, the left hand side of the Fig. 9 shows the number of joint publications between each pair of authors. The main diagonal depicts the number of single-author publications. The last three columns on the right hand side of the figure show the total number of papers, which the author published, the total number of different co-authors of the author, and the number of collaborations of the author, which is defined as the sum of joint publications of a user with each co-author.

5 Conclusions and Future Work

Within this paper, we presented AcaSoNet, a social network system for researchers with the goal of applying social network systems in a professional environment. With the help of this system, scholars gain benefit by being able to analyze their research performance with respect to the number of publications and different indexes (e.g., h-Index).

The system design of AcaSoNet finds a compromise between data reliability and scalability by combining Web mining techniques for extracting publication data with verification of the obtained publication data through the user. Since the system design is based on Web service technology, it can use external services for searching (e.g., Google Scholar) and can offer services (e.g., MyTiesTo) for providing social network information of authors. Within this paper, we also presented the user interface of AcaSoNet.

In the future, we plan to extend AcaSoNet by services for recommending publications and for automatic research group clustering based on publication lists of scholars.

References

1. Abbasi, A., Altmann, J., Hwang, J.: Evaluating scholars based on their academic collaboration activities: Two indices, the RC-Index and the CC-Index, for quantifying collaboration activities of researchers and scientific communities. In: Scientometrics. Springer, Heidelberg (2010)
2. Altmann, J., Abbasi, A., Hwang, J.: Evaluating the productivity of researchers and their Communities: The RP-Index and the CP-Index. International Journal of Computer Science and Applications 6(2) (2009) ISSN:0972-9038
3. Adamic, L.A., Adar, E.: Friends and neighbors on the Web. Social Networks 25(3), 211–230 (2003)
4. Aleman-Meza, B., Nagarajan, M., Ramakrishnan, C., Ding, L., Kolari, P., Sheth, A.P., Arpinar, I.B., Joshi, A., Finin, T.: Semantic analytics on social networks: experiences in addressing the problem of conflict of interest detection. In: WWW Conference. ACM, New York (2006)
5. Derek, J., deSolla Price: deSolla Price: Networks of scientific papers: The pattern of bibliographic references indicates the nature of the scientific research front. Science 149(3683), 510–515 (1965)
6. Goecks, J., Mynatt, E.D.: Leveraging social networks for information sharing. In: CSCW Conference (2004)
7. Golbeck, J., Hendler, J.: Inferring binary trust relationships in Web-based social networks. ACM Transactions on Internet Technology 6(4), 497–529 (2006)
8. Golbeck, J., Parsia, B.: Trust network-based filtering of aggregated claims. International Journal of Metadata, Semantics, and Ontologies (2006)
9. Heimeriks, G., Hoerlesberger, M., van den Besselaar, P.: Mapping communication and collaboration in heterogeneous research networks. Scientometrics 58(2), 391–413 (2003)
10. Jin, Y., Matsuo, Y., Ishizuka, M.: Extracting a social network among entities by web mining. In: ISWC 2006 Workshop on Web Content Mining with Human Resources (2006)
11. Kautz, H., Selman, B., Shah, M.: Referral Web: Combining social networks and collaborative filtering. Communications of the ACM 40(3), 63–65 (1997)
12. Kousha, K., Thelwall, M.: Google scholar citations and Google Web/URL citations: A multi-discipline exploratory analysis. Journal of the American Society for Information Science and Technology 58(7), 1055–1065 (2007)

13. Matsuo, Y., Mori, J., Hamasaki, M., Nishimura, T., Takeda, H., Hasida, K., Ishizuka, M.: POLYPHONET: An advanced social network extraction system from the Web. Journal of Web Semantics 5(4), 262–278 (2007)
14. Mika, P.: Flink: Semantic web technology for the extraction and analysis of social networks. Journal of Web Semantics, vol.3(2) (2005)
15. Miki, T., Nomura, S., Ishida, T.: Semantic Web Link Analysis to Discover Social Relationships in Academic Communities. In: SAINT, pp. 38–45. IEEE Computer Society, Los Alamitos (2005)
16. Mori, J., Sugiyama, T., Matsuo, Y.: Real-world oriented information sharing using social networks. In: ACM SIGGROUP Conference on Supporting Group Work (2005)
17. Nann, S., Krauss, J., Fuehres, H., Gloor, P., Fischbach, K.: Identifying Influentials by Example - the MVP (Most Valuable Player) Algorithm. Sunbelt 2010, Lago Di Garda (2010)
18. Tyler, J., Wilkinson, D., Huberman, B.: Email as spectroscopy: Automated discovery of community structure within organizations, pp. 81–96. Kluwer, B.V (2003)
19. Wasserman, S., Faust, K.: Social Network Analysis: Methods and Applications. Cambridge University Press, Cambridge (1994)

Use of Swarm Intelligence to Involve Customers in Product Innovation

Sandro Georgi and Reinhard Jung

University of St. Gallen, Institute of Information Management,
Mueller-Friedberg-Strasse 8, CH-9000 St. Gallen
{sandro.georgi,reinhard.jung}@unisg.ch

Abstract. For today's customers there is a need for integrated products and service bundles, e.g. a trip with flights, accommodation, local activities, home care during absence etc. To define such bundles, the collective intelligence of customers can be used. Thus, more customer-oriented products (bundles) can be created and the quality of bundles can be increased through the participation of many customers. By letting the collective in the form of customers participate, a solution is being created, which is better than any solution defined by a single company or any individual of the collective. To achieve this, swarm intelligence, as it can be found with social insects, is being transferred into an innovative digital platform. This platform enables customers to work together as a swarm and provides the means for customers to collectively design bundles and learn from each other.

Keywords: collective intelligence, swarm intelligence, innovation, customer orientation.

1 Introduction

1.1 High Willingness of People to Participate in Online Communities

A completely new situation in terms of availability and provision of information has resulted from the increasing penetration of everyday life with information and communication technology. Today, customers often use a large number of different communication channels for obtaining information before they come to a decision e.g for an investment; different blogs, video portals, forums, and manufacturers' websites are taken into consideration. Customers do not only search for pieces of information but get in touch with other customers and sales representatives as well. They interact in networks, exchange information on their experiences with products and share their ideas for improvements or even new products. Such available information is of great value to companies. The increase in density of crosslinking of people and companies leads to an increase in the complexity of the system in which they interact. In the so-called Web 2.0, for example, a huge amount of content is available whose quality often cannot adequately being assessed. The search for reliable information in such highly complex systems is sometimes very difficult. In the 1950s Ashby postulated a now widely accepted law of cybernetics [1], [2]), which states that the controlling

T.J. Bastiaens, U. Baumöl, and B.J. Krämer (Eds.): On Collective Intelligence, AISC 76, pp. 63–74.
springerlink.com

system of a complex system is at least as complex as the controlled system. Looking closer at this law in the context of today's complexity of markets, it soon becomes clear that it is difficult for an individual or a company to cope with this state of affairs alone. In view of the many information channels available today combined with the often unknown quality of information it can be concluded, that companies must develop new methods and tools to provide the information requested by customers. This includes new forms of cooperation and assessment of information's quality.

1.2 Research Question

In this paper the use of swarm intelligence is being discussed. In a nutshell swarm intelligence describes the behaviour of a self-organized group whose individuals act according to simple rules without a central element of coordination. As means of self-organization interaction and communication between individuals is necessary. The thus emerging swarm behaviour is many times more powerful and intelligent than isolated actions of any individual would be. Transferred to the objective of this paper, swarm behaviour of people (customers) can be fostered through communication and cooperation of several people (swarm) and the necessary provision of supporting infrastructure (swarm environment). This provided, a swarm of customers should be able to describe and create product bundles. The added value for the members of the swarm is created by the exchange with other members and the associated access to their knowledge, and by letting customers actively participate in the process of designing meaningful new products that are better tailored to their needs. Based on these facts the following research question can be derived:

> Firstly, can collective intelligence be used to let customers participate in the innovation process, particularly in the design of product bundles, and secondly, by what means can this be enabled?

The two questions are addressed in this paper as follows. In the second chapter the theoretical framework of swarm intelligence is being described. Different definitions of a swarm and how they can be distinguished are presented, and the key factors for swarm intelligence are compiled. In chapter 3, the applicable theory to solve the first research question is discussed and the characteristics of an appropriate digital swarm environment in the form of a prototype are shown. In the fourth chapter, initial test results are presented and the prototype is being evaluated with regard to how well the problem posed by the first research question is being solved. Chapter 5 contains a summary of the results and an outlook on further research.

1.3 Research Approach

In the field of economics and social sciences only few studies on swarm intelligence can be found, as well as is in the area of information system [3], [4], [5]. In the field of biology, are, for obvious reasons, several different studies available [6], [7], [8].

The transfer of the swarm mechanisms from biology to a group of people and the evaluation of the transformation by using a software prototype is assigned to the design-science-research approach used in the field of information systems. This allows for obtaining evidence concerning the behaviour of customers and reviewing the defined characteristics of a swarm platform. This work is set within the context of design-research [9], which purpose is the development of a meaningful artefact, which solves an existing problem. The scientific approach of behavioural-science which is the "(reactive) analysis of design and the impact of available IT artefacts on companies and markets" [10] is not pursued here.

2 The Concept of Swarm Intelligence

In biology, a swarm is generally regarded as a decentralized and self-organized group of similar animals [11]. A narrower description of the concept of a swarm assumes that individuals are homogeneous, i.e. that they are of the same species [12]. Typical examples of such animals are ants, bees, termites and other social insects [9]. In this paper the narrower swarm concept is being used.

2.1 Successful Swarms Consist of Many Individuals

According to Bonabeau and Meyer [3], and Dorigo and Birattari [12] the following typical characteristics are assigned to swarms. (1) A swarm consists of many individuals: except for Kazadi [13] no exact numbers can be found in literature on how many individuals are needed to build a swarm. Kazadi mentions that a swarm consists of at least two or more individuals; however, against the background that a minimum flock size is necessary for the emergence of a sufficiently strong pheromone[1] trail in search of food, this low limit does not seem to be appropriate. Therefore, a swarm is considered to be composed of many individuals; this means considerably more than two or three individuals. (2) The individuals of the swarm are homogeneous, which means that they are identical (e.g. robot swarms), or all of the same species (animals). In nature, individuals feature, however, even if they are of the same species, different sets of genes, have different experiences and slightly different behavioural patterns and are therefore heterogeneous regarding their behaviour [6], its own experiences and diverse set of genes. (3) Individuals of a swarm are, compared with the abilities of the swarm as a whole, of limited intelligence, and posses fewer capabilities. The swarm as a whole is many times more powerful than the single individual. Individuals in a swarm act according to simple (swarm) rules [3]. These rules create a sometimes completely unexpected behaviour; e.g. in a flock of birds each individual stays within a certain distance to its neighbouring bird and flies slightly to the rear of it. By following these two simple rules, the familiar V-shape of bird flocks emerges. Similar behaviour can also be observed for schools of fish [14]. Systems, which behave in such a way, can best characterized as emergent systems. It is not possible to deduce the behaviour of the wholes system from the behaviour of the subsystems [15]. The peculiarity of emergence renders swarms particularly interesting for research.

---

[1] Pheromones are chemical messengers secreted by ants during food transport to the nest. Conspecifics olfactorily perceive these substances and use them as markers on their way to a food source. On the way back to the nest, they also lay down pheromone, and reinforce the track. At the same time pheromone evaporates over time. Thus a trail less trodden remains a weak pheromone track and is less likely to be followed than a strong trail [3], [6], [16].

Bonabeau and Meyer [3] state three main reasons why swarms are as highly successful as they are: (1) flexibility, (2) robustness and (3) self-organization. (1) A swarm has the flexibility to quickly adapt to a changing environment. Ants are able find a new location for their anthill very quickly if the old one is compromised or destroyed [6], [7]. (2) Due to the large number of its individuals a swarm is robust. The swarm as a whole continues to exist, even if single individuals fail or perish. There are numerous other animals, which can take over the work (3). The most important aspect for success mentioned by Bonabeau and Meyer is self-organization. Ants act independently, i.e. without central coordination, and base their decisions only on pieces of information locally available.

2.2 Swarms are Self-organized and Communicate Stigmergically

Four conditions [11] must be met for a swarm to be self-organized: (1) positive and (2) negative feedback, (3) fluctuations in behaviour, and (4) the opportunity for numerous interactions must exist. (1) If ants lay a pheromone trail, they do so to motivate their peers to follow the trail to a source of food and to reinforce the existing trail. This amplification mechanism is called positive feedback, as this increases and reinforces an existing behaviour (folling a trail) [6]. (2) While positive feedback multiplies behaviour, it can be limited by negative feedback. A food source can be located so unfavourably that it may be subject to exploitation by only a limited number of animals [6]. The same applies to a pheromone trail that dissipates by evaporation when it is trodden at a low rate or if the exploitation of a food source is stopped because the source is depleted. A trail which is no longer used and on which no more pheromone is lain down completely evaporates within a certain period of time. Communicating with each other with pheromones ants change their environment (laying down pheromone). This form of indirect communication is known as stigmergy [17] where one animal (e.g. an ant) changes its environment and other animals react to it. (3) The behaviour of ants on following an existing pheromone trail is up to a certain degree of random nature. This allows for new food sources to be discovered. If ants would always follow existing pheromone trails or would never ignore an existing trail, no new food sources would be discovered in sufficient quantity. Fluctuation in the behaviour of animals and the associated randomness are typical characteristics for swarms and necessary for self-organization [11]. (4) A swarm is also characterized by numerous interactions among its individuals. The exchange of information takes place between only two or at the most a few individuals. This requires a large number of interactions to help disseminate information and knowledge [11].

In this chapter we have shown that swarms consist of many homogenous individuals with heterogeneous behaviour. If these individuals act together as a swarm following simple (swarm) rules the behaviour of the swarm is much more powerful than the sum of the individuals' behaviour. The most interesting point concerning swarms is their ability for self-organisation. This entails positive and negative feedback mechanisms, fluctuation in behaviour and numerous interactions. The fascinating phenomenon of swarm intelligence can now be applied to the cooperation of customers. How this can be achieved will be demonstrated in the next chapter using a concrete example.

3 Application of Swarm Intelligence for the Definition of Product Bundles

Nowadays various phenomena can be observed where customers get involved in product innovation and/or sharing ideas and knowledge with each other and companies. This willingness to engage with and assist in the development of products serves customers and companies alike. Companies learn more about their customers' needs while customers get products which better suit their needs. The concept of swarm intelligence as described in the previous chapter is an innovative way to bring customers and companies together. How this can be achieved is explained using the example of bundles consisting of integrated products and services.

3.1 Customers Need Integrated Products and Services

When analyzing current customer needs, it can be stated that these often no longer consist of just one product or service, but of a number of products and services that are combined into an overall package [18] and as such expand into larger customer processes. The required integration of the individual products into a coordinated bundle of products and/or services is usually done by the customers themselves. The resulting bundle is a combination of products and/or services from different vendors [18], [19], and may be, for example, the realisation of a journey. The main focus of such bundles, therefore, is on integration of and not on individual products and services. Anderson [20] refers to this kind of bundle as a complete offering which consists of products and services and leaves no additional work to be done for the customer (to which we refer as "integrated"). According to Goldman et al. [21] an offering which consists of products and services can be transferred into a solution which satisfies the needs of the customer. Such solutions are much more valuable than a number of isolated products and service, which must be integrated by the customer himself. We therefore define a product bundle as consisting of products and/or services that fully satisfy the specific requirements of a customer (process) while there is no additional work left to be done for the customer. Friedman and Langlinas emphasize that the value provided for the customer for such integrated products and services is that the value of a bundle is greater than the sum of the parts [19]. Proposals concerning the composition of such bundles should be made by the customers themselves (as they know their own needs best), and should be made available to other customers who can learn from this knowledge, when creating their own bundles. Ideally, people cooperate with each other and can achieve product bundles of a higher quality. Thus, they can learn from each other and get integrated products and/or services by which the coordination effort is omitted for customers, e.g. done by a service integrator[2].

[2] In this paper the the main focus lies on the point of view of the service integrator who provides a single point of contact for customers for creating bundles of products and/or services. Products and services are considered as goods which are "ordered" by a customer and integrated by the service integrator. As such both products and services are treated the same without the main distinction of the so-called customer-integrating process [23]. For the sake of simplicity "product" is used as a synonym for "product and/or services" in the future course.

3.2 Swarm Behaviour Adapted to Create Product Bundles

There is a multitude of parallels between 'animal' and 'human' swarms. While a swarm in nature is constituted by individuals of one species, one in the business world consists of customers e.g. from one company. While ants live in their natural habitat and optimize the trails leading to food, customers need a digital environment in form of an online application to act as swarm and optimize product bundles. The pieces of information produced while a customer chooses products and includes them into his bundle can be used to show customers how other customers' bundles look like. A product bundle can be represented as a directed graph with a source (*start*) and a sink (*end*). The sequence of products corresponds to the order of consumption of the products. In this way, each product has at least one predecessor (*start* or other products) and one successor (other products or the *end*). Between each two nodes a directed edge shows the sequence of consumption. Every time a product is chosen as successor to another product, the amount of digital pheromone [22] for this particular product combination is increased by one unit. Like natural pheromone digital pheromone also evaporates over time. The evaporation rate can be calculated with a half-life function. Thus for every product and its successors the amount of pheromone can be deduced. The higher a concentration of pheromone is, the more frequently this particular product combination was chosen.

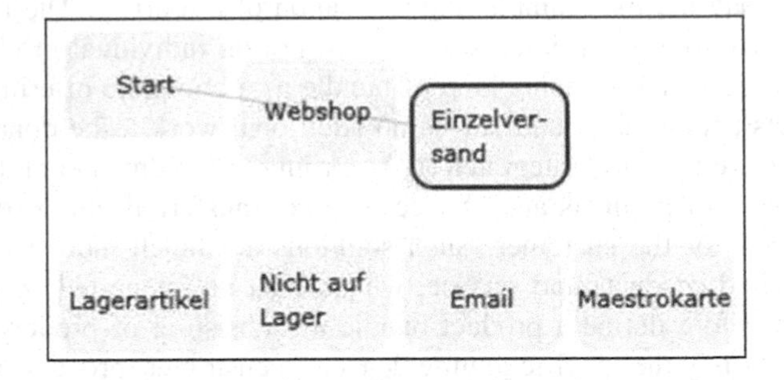

Fig. 1. Directed Graph of Products and Successor Products to "Einzelversand"

If several of these product combinations are strung together (Fig. 1) a trail similar to an ant trail can be detected. While ants choose their next steps literally because of the amount of pheromone found at their current location[3], a customer "A" can be displayed what products were chosen by other customers who included the same product as customer "A" just included. Customer "A" can now choose either from these suggestions or from a regular catalogue which contains all available products. In Fig. 1 the last product chosen was "Einzelversand" ("ship each good as soon as it is available"). The products displayed below the graph are the products other customers have chosen as

[3] Ants base their decision on which trail to follow solely on locally available pieces of information (pheromones). Compared with customers putting together product bundles, this means, that any available information (digitale pheromone) must transparently be displayed for everyone [24].

successors to "Einzelversand". Their order from left to right corresponds to the amount of pheromone (how often these products were chosen over time). With the use of digital pheromone and its accumulation and evaporation a mechanism for positive and negative feedback is provided. Positive and negative feedback are alongside with fluctuation in behaviour (not always choosing from the suggested successors but also from the regular catalogue) and frequent interactions (perceiving pheromone lain down by other customers) the necessary prerequisites for self-organization in a swarm. To enhance the possibility of fluctuation, every customer can create new products that he finds missing in the existing catalogue. These new products can be incorporated by all other customers into their bundles. If a customer creates a new product and this particular new product is subsequently used a number of times by other customers, a new and strong pheromone trail for this product combination is created and gives evidence to new products needed.

In this chapter we introduced the concept of integrated product bundles and we have shown how such bundles can be created by customers acting as a swarm (working together) and learning from each other by means of pheromone trails. In the following chapter we show how this swarm mechanism can be supported by an appropriate (digital) environment and describe a prototype of such an environment.

4 Prototype

To enable customers to act as a swarm an appropriate swarm environment is required. The implementation of such an environment in the form of a prototype and initial test results with the prototype are being described in this chapter.

4.1 Prototype

For the development of the prototype two main priorities have been pursued. On the one hand an easy to use and intuitive interface design was built and secondly the underlying logic of swarm intelligence as described above was implemented. The graphical user interface (GUI) was constructed by analogy to the Lego Digital Designer (LDD) [25]. The LDD was constructed for children in order to digitally assemble Lego toys, to share the designs with other children over the internet[4] and to order the used Lego bricks online. The GUI is very easy to use and consists of four main areas: there is a header with a menu, a toolbox that contains all available Lego bricks, a contextual menu for the use of the bricks and the work space where the Lego toy is being assembled.

The GUI[5] of the swarm prototype consists of the same four areas (Fig. 2). At the very top (1) is the menu (for saving, deleting etc.) for the whole application. Located on the left hand side is the product catalogue (2) which is divided into three areas; at the top is the whole catalogue, in the middle all products created by customers are

[4] In this context Lego uses a similar approach for the use of collective intelligence as it is proposed by Gloor and Cooper [5], [24]. Lego does not follow nor apply the strict swarm mechanisms explained in this paper.

[5] For constructing the GUI the laws from Ehrenfels [26] and EN ISO 9241 [27] have been followed.

displayed and at the bottom only products created by the particular customer are visible. Located at the bottom of the screen is a contextual menu, which shows information and functions for the product selected on the work space (4).

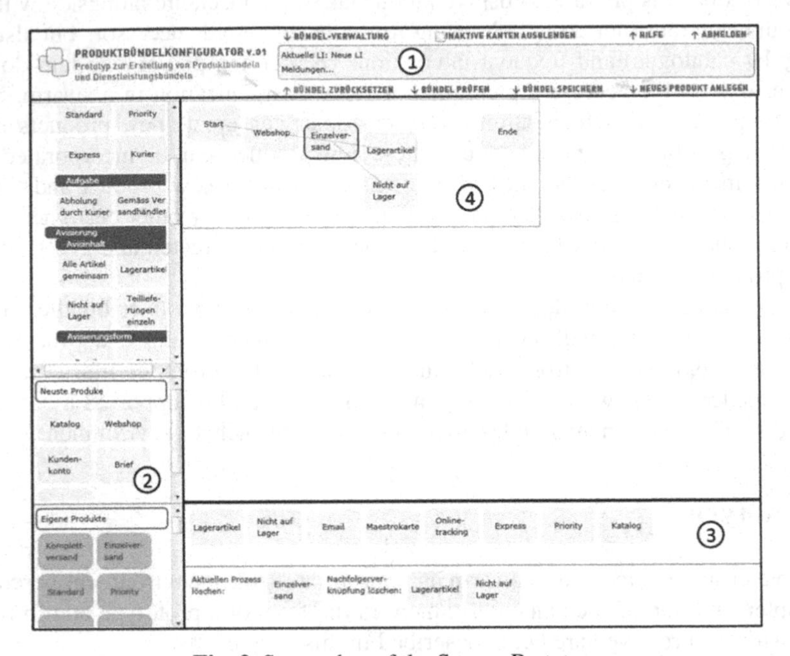

Fig. 2. Screenshot of the Swarm Prototype

Product bundles can be put together in different ways. To add a new product, the predecessor of the new product is selected by a mouse click. In the contextual menu the costumer is shown which other products have been chosen as successor to the selected product. These recommendations are based on the level of pheromone for each product combination (consisting of a predecessor and a successor). The product with the highest level is displayed on the far left side. Thus presented products can be added by clicking on them. Alternatively the customer can also choose from all three parts of the product catalogue by clicking on one of those products.

Regardless of a product's origin as successor to an already added product, the level of pheromone for this particular product combination is increased by one unit. This piece of information disseminated to all other customers in the form of recommendations for product combinations. By doing so, positive and negative[6] feedback mechanisms are established. Furthermore, customers interact by using these pieces of information and by creating and using new products. Positive and negative feedback, interaction and communication (realised by displaying important information to all customers) are the necessary parts of swarm intelligence and its self-organisation.

A good example of a simple rule which has to be followed by the members of the swarm is, that every product in the bundle can have several successors. Thus created

[6] Evaporation (~weakening of the trail) of pheromone is continuously calculated by a half-life function as described in chapter 3.

branches are considered to be logical AND-connections. This means, that if one product has two successors, the two successors must be "consumed" simultaneously and consumption must be finished before the two branches are conjoined again.

4.2 Evaluation of the Prototype

In order to verify the developed prototype two tests have been carried out so far. Both tests were conducted in lab-like situations with university students who were given the task to define their favoured process for online shopping (e.g. which steps in which order they would prefer to go through, choosing preferred payment options, shipping options etc.). The tests were carried out with information systems students, who were familiar with using digital design tools and online platforms. As such there are no unbiased conclusions possible concerning the usability of the user interface. The main focus therefore was on whether swarm behaviour occurs or not. Furthermore, the participants had no real life decisions to make. Their participating could best be described as noncommittal. To start the tests the students were given a brief introduction on how to use the prototype and were then set to work. In order not to influence the participants' behaviour no explanations were given on the subject of swarm intelligence. The first test took place over a period of time of thirty minutes (with a half-life value of 6 hours) while the second test lasted twelve hours (with the same half-life value as for the first test). The first test group consisted of 17 persons, the second group of 30 persons.

To determine whether the desired swarm behavior has occurred through the use of the prototype, the following metrics have been developed. Swarm behavior may occur by (1) the use of recommended and latest products, (2) the creation of new products and the course of their use and (3) the development of unique pheromone traces of predecessor-successor product combinations.

When creating product bundles customers can choose either products from the catalog (~ non swarm element) that do not foster swarm behavior or products from recommendation or the latest products (~ swarm elements), that enhance swarm behavior. Using this distinction makes it possible to determine the frequency of the choice of swarm and non swarm elements as a percentage value of all product selections (the amount of selections amounts to 100 percent). As no such experiment has yet been conducted, no data is available to what extent swarm elements should be used. Based on the resulting experiences from the work with the prototype, the author expected, however, a percentage of use of swarm elements in the range of 20 percent.

Both tests showed an even higher usage of the swarm elements (Table 1) which is preliminary considered to be a good result.

Table 1. Distribution of the Use of Swarm and Non Swarm Elements

	Test 1		Test 2	
	Clicks	Percent	Clicks	Percent
Swarm elements	170	36%	112	26%
Non swarm elements	306	64%	318	74%
Total	476	100%	430	100%

It was possible to discern in both tests strong pheromone trails for certain product combinations and to filter out some clear customer preferences for the composition of the required product bundle.

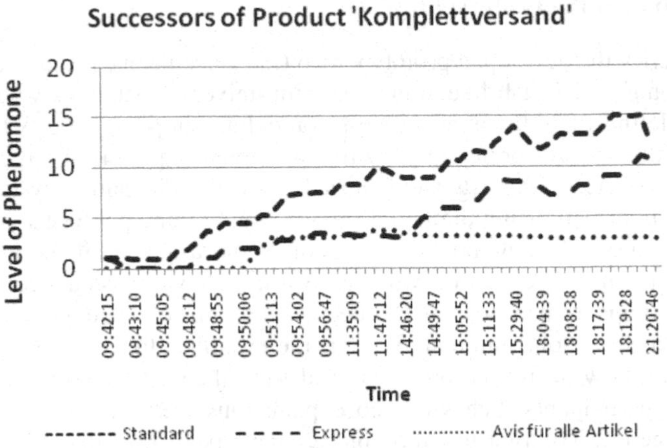

Fig. 3. Successors of Product "Komplettversand"

In Fig. 3 the level of pheromone (~pheromone trails) of three successors of the product "Komplettversand" ("group all goods together for shipping") is displayed. While the level for product "Avis fuer alle Artikel" ("shipping notice") remains low throughout the test, significantly higher levels of pheromone can be detected for the other two products ("standard shipping rate" and "one-day shipping rate"). With the help of these pheromone trails it is possible to discern, that a majority of the test persons prefer the standard shipping rate while less would want "one-day shipping". Only very few prefer to receive a shipping notice. In the light of shipping goods, sold in an online store, this insight might seem trivial. But the potential for companies having their customers define their preferred process of whatever transaction thinkable, in combination with receiving the information what the process preferred by the majority of customers looks like is enormous. This is also what distinguishes the proposed method from association rules for the evaluation of shopping baskets as proposed by Agrawal et al. [28]. Association rules typically explore which goods are bought together. But there are no conclusions on the sequence of the order of goods. In order to provide ideal customer processes it is indispensable to know the order in which customers intend to "consume" the products (the chronological sequence of the process). Using pheromone trails this kind of knowledge is made available while at the same time the strongest pheromone trail represents the most favoured customer process. Properly applied this can help companies (service integrators) to vastly improve their product bundles and increase their focus towards customers' needs. Over all, those first test results presented are very promising but the prototype must be refined and further evaluated.

5 Outlook

We raised the question, whether collective intelligence could be deployed to let customers participate in the innovation process, particularly in the design of product bundles, and in what form this could be achieved. The objective of the paper has been achieved, i.e. to demonstrate that collective intelligence can be successfully applied to the definition of product bundles. Further research must now be conducted to improve the prototype and refine the underlying swarm mechanisms. The comparison of pheromone levels increasing over time with the typical product lifecycle might show interesting similarities and patterns respectively. The GUI must be improved in order to locate recommendations in a more aggressive manner, so they get better noticed and thus the amount of usage of swarm elements can be raised. It should also be investigated, what kind of expectations customers have of a company for voluntarily providing their knowledge. Does it already suffice to provide only a suitable and innovative platform or do customers expect financial compensation for their work or is the attention and recognition for one's contribution in such a network key?

References

1. Ashby, W.R.: An Introduction to Cybernetics. Wiley, New York (1956)
2. Heylighen, F.: Principles of Systems and Cybernetics - an Evolutionary Perspective. In: Cybernetics and Systems 1992, pp. 3–10. World Science, Singapore (1992)
3. Bonabeau, E., Meyer, C.: Swarm Intelligence - A Whole New Way to Think About Business. Harvard Business Review, 107–114 (May 2001)
4. Tarasewich, P., McMullen, P.R.: Swarm Intelligence – Power in Numbers. Communications of the ACM 45, 62–67 (2002)
5. Gloor, P.A., Cooper, S.M.: The New Principles of a Swarm Business. MIT Sloan Management Review 48, 81–84 (2007)
6. Sumpter, D.J.T.: The Principles of Collective Animal Behaviour. Philosophical Transactions of the Royal Society of London 365, 5–22 (2006)
7. Kelly, K.: Out of Controll – The New Biology of Machines. Fourth Estate, New York (1994)
8. Couzin, I.D., Krause, J., Franks, N.R., Levin, S.A.: Effective Leadership and Decision Making in Animal Groups on the Move. Nature 433, 513–516 (2005)
9. Hevner, A.R., March, S.T., Park, J., Ram, S.: Design Science in Information Systems Research. MIS Quarterly 28, 75–105 (2004)
10. Hess, T., Wilde, T.: Forschungsmethoden der Wirtschaftsinformatik: eine empirische Untersuchung. Wirschaftsinformatik 49, 280–287 (2007)
11. Bonabeau, E., Dorigo, M., Theraulaz, G.: Swarm Intelligence – From Natural to Artificial Systems. Oxford University Press, New York (1999)
12. Dorigo, M., Birattari, M.: Swarm Intelligence. Scholarpedia 9, 1462 (2007)
13. Kazadi, S.: Swarm Engineering. PhD-Thesis at the California Institute of Technology, Pasadena (2000)
14. Couzin, I.D., Krause, J., James, R., Ruxton, G.D., Franks, N.R.: Collective Memory and spatial Sorting in Animal Groups. Journal of Theoretical Biology 218, 1–11 (2002)
15. Mayr, E.: Das ist Biologie – Die Wissenschaft des Lebens. Spektrum Akademischer Verlag, Berlin (2000)

16. Beekmann, M., Sumpter, D.J.T., Ratnieks, F.L.W.: Phase Transition Between Disordered and Ordered Foraging in Pharaoh's Ants. Proceedings of the National Academy of Sciences of the United States of America 98, 9703–9706 (2001)
17. Grassé, P.-P.: La réconstruction du nid et les coordinations interindividuelles ches Bellicositermes Natalensis et Cubitermes sp. La théorie de la Stigmergie: Essai d'interprétation du comportement des termites constructeurs. Insectes Sociaux 6, 41–81 (1959)
18. Baumöl, U., Winter, R.: Intentions Value Network: a Business Model of the Information Age. In: Proceedings of the Third International Conference on Enterprise Information Systems, pp. 1075–1080. ICEIS Press, Setubal (2001)
19. Friedman, J.P., Langlinas, T.C.: Best Intentions: a Business Model for the eEconomy. Outlook 1, 34–41 (1999)
20. Lastminute.de, `http://www.lastminute.de`
21. Expedia.de, `http://www.expedia.de`
22. Aberer, K., Wu, J.: Swarm Intelligent Surfing in the Web. In: Cueva Lovelle, J.M., González Rodríguez, B.M., Joyanes Aguilar, L., Labra Gayo, J.E., de Ruiz, M. (eds.) ICWE 2003. LNCS, vol. 2722, pp. 431–440. Springer, Heidelberg (2003)
23. Kleinaltenkamp, M.: Integrativität als Kern einer umfassenden Leistungslehre. In: Backhaus, K., Günter, B., Kleinaltenkamp, M. (eds.) Marktleistung und Wettbewerb, Gabler, Wiesbaden (1997)
24. Gloor, P.A.: Swarm Creativity - Competitive Advantage through Collaborative Innovation Networks. Oxford University Press, New York (2006)
25. Lego Digital Designer, `http://ldd.lego.com/`
26. von Ehrenfels, C.: Über Gestaltqualitäten. Vierteljahresschrift für wissenschaftliche Philosophie 14, 249–292 (1890)
27. EN ISO 9241, `http://www.iso.org`
28. Agrawal, R., Imielinski, T., Swami, A.: Mining Association Rules Between Sets of Items in Large Databases. In: SIGMOD 1993, pp. 207–216 (1993)

Imitation and Quality of Tags in Social Bookmarking Systems – Collective Intelligence Leading to Folksonomies

Fabian Floeck, Johannes Putzke, Sabrina Steinfels, Kai Fischbach, and Detlef Schoder

University of Cologne, Department of Information Systems and Information Management, Pohligstr. 1, 50969 Cologne (Köln), Germany
{Putzke,Steinfels,Fischbach,Schoder}@wim.uni-koeln.de

Abstract. Social bookmarking platforms often allow users to see a list of tags that have been used previously for the webpage they are currently bookmarking, and from which they can select. In this paper, the authors analyze the influences of this feature on the tag categorizations resulting from the collaborative tagging effort. The main research goal is to show how the interface design of social bookmarking systems can influence the quality of the collective output of their users. Findings from a joint research project with the largest Russian social bookmarking site *BobrDobr.ru* suggest that if social bookmarking systems allow users to view the most popular tags, the overall variation of keywords used that are assigned to websites by all users decreases.

Keywords: Collective Intelligence, Collaborative Tagging, Folksonomies, Shared Knowledge, Social-Bookmarking-Systems.

1 Introduction

In recent years, social bookmarking services such as *BobrDobr.ru*, *citeulike.org*, *del.icio.us*, and *mister-wong.de* have gained an increasingly large user base [e.g. 1]. By using a social bookmarking service, users can bookmark objects on the World Wide Web – identified by their Unified Resource Locators (URLs) – and annotate each object with metadata, or so-called "tags". A tag is a keyword that describes the annotated object from the user's point of view. The process of many users assigning arbitrary tags to shared objects is often called "collaborative tagging," and the set of tags that results is typically denoted "folksonomy" [2, 3].[1]

Considerable research has been devoted to the suitability of folksonomies for content classification, and particularly to the tradeoff between the users' bottom-up approach of assigning free keywords for classification and the quality of top-down-defined classifications created by experts [e.g. 5]. However, rather less attention has

[1] Since the term "folksonomy is said to have been coined by Vander Wal, the authors reference some of Vander Wal's blog entries. However, there are many other (academic) definitions of folksonomies and their characteristics [e.g. 4].

T.J. Bastiaens, U. Baumöl, and B.J. Krämer (Eds.): On Collective Intelligence, AISC 76, pp. 75–91.
springerlink.com

been paid to whether the quality of folksonomies depends on the functionalities offered by the bookmarking system used to create them. In this paper, the authors analyze empirically the influence of functionalities that allow users to see which tags others have assigned to a URL during the process of bookmarking on the resulting set of tags of these URLs.

The remainder of this paper is structured as follows. Section 2, Related Work, reviews the literature that examines the quality of folksonomies and their suitability for content classification, as well as works that deal with the characteristic and effects of imitation. Through this, possible scenarios for the effects of imitation in a social bookmarking system are drawn. In Section 3, Hypotheses, the authors develop four hypotheses to test the impact of the visibility of the most popular tags for a certain URL on the variation of tags assigned by the users for this URL. Section 4, Analyses, describes the dataset, the methodology, and the operationalization of variables. In Results (Section 5), the authors describe the procedures used for the empirical analyses and highlight the findings. Finally, Section 6, Summary and Conclusions, discusses the implications of the findings, notes their limitations, and provides some suggestions for further research.

2 Related Work

Folksonomies are one of the current research trends in a variety of academic disciplines such as information systems [e.g. 6, 7], computer science [e.g. 8, 9], physics [e.g. 10, 11], anthropology, and sociology [e.g. 12]. Hence, a search for "social bookmarking", "social bookmark", "folksonomy" or "social tagging" in the title / keywords or abstract fields in "Web of Science"[2] yielded 306 articles published between 2005 and 2009.[3] Since it is not possible to review the entire related literature here, the literature review will focus on studies that examine the emergence of folksonomies in social bookmarking systems [e.g. 13]. Before taking this specific focus, the authors also provide some insights into the applications and uses of social bookmarking services.

There are different potential usage scenarios for tags and social bookmarking services. The most evident applications might be the use of folksonomies for web search optimization [e.g. 14, 15] and knowledge organization [e.g. 16], since folksonomies can facilitate detection of non-explicit properties of web objects. For example, a user can characterize a web page by annotating it with the tag "funny" [13]. An automated system that extracts existing metadata from web pages cannot achieve this, as the system cannot comprehend this property. Therefore, folksonomies might also be a useful approach for the semantic web [e.g. 17].

Social bookmarking services are typically free of charge and require little knowledge to use. This increases active participation by many users [18], who not only manage their own bookmarks with social bookmarking services, but also use the service to search in other user's public bookmarks for web objects with specific properties, represented by tags that serve as a filter. This allows all users of the service to

---

[2] http://apps.isiknowledge.com/.

[3] Note that this keyword search did not include the terms "ontology" or "classification".

benefit from all other bookmarks [19] in addition to the individual benefits of users from their own bookmarks [18].

Folksonomies resemble a "bottom-up" approach for generating a vocabulary. Some researchers argue that tags, unlike controlled vocabularies or hierarchical taxonomies, more accurately reflect users' conceptual models. The idea is that every user adds his or her individual conceptual model of a piece of content to the pool of tag descriptions, which in turn is accounted for in the aggregated description of the content. Thus in social bookmarking systems, a large number of users – each investing their own cognitive resources – is collectively making sense of specific web contents. The result might be considered a "democratically agreed upon" description of content.

For the social bookmarking system *del.icio.us*, Golder and Huberman [13] found that after a certain number of bookmarks has been created for a specific URL, a stable power law pattern with fixed proportions of frequency for each tag can be observed, no matter how many bookmarks are stored for that particular URL after that point. They relate the emergence of this stable pattern to two basic explanations.

The first explanation is "shared knowledge". Golder and Huberman [13] argue that sharing the same experiences that may be universal within a culture or community leads to similar ways of sensemaking. As a consequence, categories emerge that are widely agreed upon, co-existing with personal categories that are rarely reproduced. Accordingly, experiments showed that basic-level categories are the most probable classifications made when objects are perceived for the first time [20], although an individual's expertise in a specific field influences what she or he considers to be a basic category [21]. As the variation of basic-level categories is lower the more general they are, and more general descriptions are preferred, users agree on general descriptions but differ on more specific ones. Other researchers trace the users' behavior of preferring broad and simple categories over specific ones back to the aim of investing the "least cognitive effort" [e.g. 4]. Thus, following the idea of a shared conceptual model, the power law tag pattern for given content may, over time and with more users bookmarking, come closer to a realistic representation of what the user base collectively "thinks" that content is about.

The second explanation for this pattern is the presence of *imitation* behaviour. Some tags (e.g., basic categories) are popular; by being imitated, they become even more popular over time, eventually forming and amplifying the power law pattern. Imitation behaviour can also be traced back to the need to save cognitive ressources [4] and is facilitated by a feature common to many bookmarking systems that makes it possible to see the most popular tags assigned to a URL. Users are looking for the best tag choices without having to think too much about them, and they are often indecisive about the right tagging choice. Therefore, they trust the "social proof" [22] offered by the decisions of the majority. It is not a wish to conform that explains this "rational imitation" [23], but rather the need of the individual user to arrive at a better tagging decision.

What results when users "follow the behaviour of the preceding individual" [24] or individuals is an "information cascade". With every imitated decision, the individual puts aside her or his individual tagging choices and the stability of the information cascade grows as the popular tags (e.g., basic categories) become even more popular. As Bikhchandani et al. [24, pp. 1006, 1009] put it: "Intuitively, cascades aggregate the information of only a few early individuals' actions." The result: "The social cost of cascades is that the benefit of diverse information sources is lost." This means that only the conceptual models of a few early individuals are the basis of the emerging collective tag description. Individual viewpoints expressed or mistakes made in the early bookmarks are consequently represented at a disproportionately high rate in the eventual distribution.

Note, however, that an information cascade in a social bookmarking system can take two forms, the latter one being different from the described process:

1. Replacement of individual tags by popular tags: the overall tag variation for a URL decreases because users do not think about tags of their own or neglect their own tag choices in favor of the tag choices of the majority. Diverse information is lost.
2. Extension of individual tags by popular tags: When opting for a popular tag or set of tags, users still assign their own full sets of tags to the URL. Here, no information is lost; there is, however, an overrepresentation of the early assigned tags that effects, for example, the display of tag clouds for navigation purposes.

These two scenarios (see Fig. 1) are not mutually exclusive but rather occur simultaneously and interfere with each other even in one single bookmarking process.

The goal of this study was to find empirical evidence 1) that there is overrepresentation of early-assigned tags because they are imitated, and 2) that this imitation leads to a loss of information in the collective description because early tags suppress and eliminate individual tags.

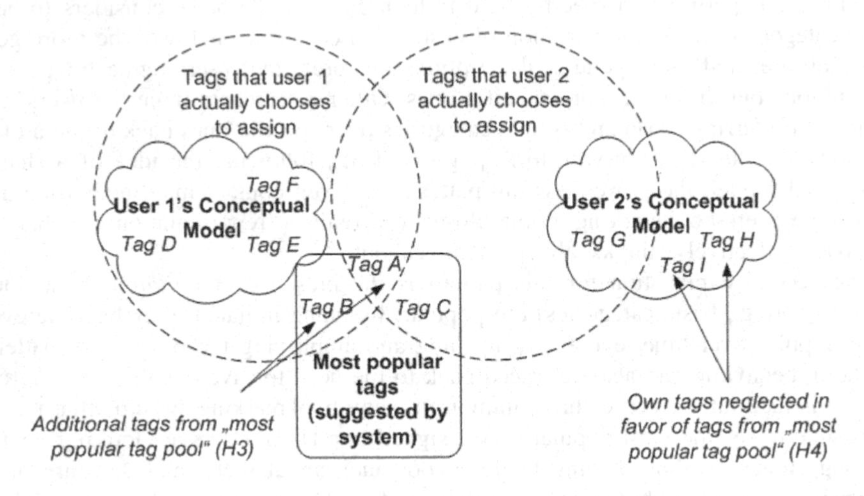

Fig. 1. Tag suggestions leading to more tags (H3) vs. suggestions leading to less tags (H4)

3 Hypotheses

Looking at the empirical findings and knowledge regarding shared categories highlighted in the previous section, the authors assume that there must exist categories or tags that users share and agree upon widely, as well as some they agree upon less – and the less they agree upon these, the more specific the tags become. The authors further assume that the tags that the users share and agree upon, therefore, will not only be assigned more often, but also will have a high probability of appearing quite early (compare [13]).[4] Hence,

***H1**: There is a positive correlation between the time when a tag is first assigned to a URL and its overall frequency for that particular URL.*

The authors suppose that ***H1*** is true whether or not users imitate other users' tags. However, the effect postulated in hypothesis ***H1*** should be stronger when imitation is made possible by the user interface. Users then agree upon the same tags, but also, tags that are assigned frequently and early already are imitated because they are suggested as "most popular". As pointed out above, it is likely that an information cascade will ensue, resulting in a higher copy rate and thus an eventual overrepresentation of early-assigned tags. Therefore,

***H2**: The positive correlation between the time when a tag is first assigned to a URL and its overall frequency for this URL is stronger if users can see tags already assigned by other users.*

Under the assumption that different users have different associations regarding a URL, it is very likely that the most popular tags contain tags a user had not thought of when he or she decided to bookmark a URL. On average, then, the user has a wider choice of tags when the most popular tags are displayed than without that functionality. The user may decide to choose some of the displayed tags, sometimes in addition to and sometimes instead of his or her own tags, as discussed in the previous section. Overall, users should assign more tags to a URL when they have the option to imitate (see Fig. 1). Hence,

***H3**: A URL receives more tags from a single user if that user can see other users' previously assigned tags.*

A user provided with the most popular tags might, according to the concepts of the least cognitive effort and social proof, neglect some of his or her own conceptual models and the associated tags. Instead, the user will settle for some of the offered popular tags and regard them as a good choice. This leads to a lower degree of variation in the set of tags a URL receives overall. Therefore,

***H4**: A URL receives fewer different keywords in the form of tags overall if that user can see other users' previously assigned tags.*

In the next section, we describe how we tested these hypotheses empirically based on the complete dataset of a social bookmarking service.

4 Analyses

4.1 Dataset

To test the proposed hypotheses, the authors conducted a joint research project with the Russian social bookmarking platform *BobrDobr.ru*, which was selected as the data source for four main reasons:

[4] This may sound trivial, but it is the basic correlation that must be shown to prove the assumptions the authors make for the imitation model.

First, *BobrDobr.ru* offers users two different possibilities for adding a bookmark to the data base: one allowing for imitation and one that leaves bookmarking up to users themselves, individually. These different procedures were crucial for the research design of this study. The two bookmarking methods work as follows. The first ("internal") method enables the user to copy a bookmark for a certain URL that has already been bookmarked by other users within the system. When a user chooses to copy and add the bookmark to her or his own library, the system displays to the bookmarking user the five tags used most often (to date) for that URL. With the second ("external") method, the user can either click on a special button installed in the web browser or a *BobrDobr.ru*-Link implemented by the website itself to bookmark the URL the user is currently visiting outside the system. No tags of other users are shown if URLs are bookmarked "externally" via the browser button. This setting allowed the authors to compare the sets of tags a URL had received from both methods: one set resulting from the "internal" method, that is, imitation of the most popular tags; and the other set resulting from the "external" method, which gave no insight into popular tags.[5]

Second, the authors wanted to analyze a social bookmarking service that is visited primarily by users from one language area only (unlike, for example, *del.icio.us*). *BobrDobr.ru* is the leading Russian social bookmarking platform[6] and has a relatively homogenous user base in terms of language due to the Cyrillic alphabet. According to the operators of *BobrDobr.ru*, 90 percent of users reside in Eastern Europe and 80 percent using a Russian system.

Third, except than the language, the platform has neither a specific target user group nor restrictions in terms of content (like, for example, *citeulike.org*).

Fourth, the operators of *BobrDobr.ru* were willing to conduct a joint research project and provided the researchers full access to their database. Furthermore, the operators allowed the researchers to determine which data they would like collected, and the operators stored these data persistently for the duration of the research (e.g. the time stamps for each action).

The data collection began on the date of the relaunch of *BobrDobr.ru* on 1 May 2008, when the two different bookmarking procedures ("external" and "internal") were introduced, and ended on 20 August 2008. The dataset contained logs for all user activity during the observation period. At the end of the data collection period, the platform had about 61,000 registered users who created about 4,000 bookmarks per day. As the authors only analyzed URLs that were first bookmarked after the relaunch date, they were able to observe the development of tag sets assigned to URLs right from the first bookmark they ever received and still analyze a large system with many users. Before analyzing the imitation of tags, the researchers also had to account for the fact that *BobrDobr.ru* offers users the option of saving bookmarks as either "public" or "private", the latter resulting in saved bookmarks not being displayed to other users nor being included among the "most popular tags" feature.

---

[5] Users of *BobrDobr.ru* also have a third option to add a bookmark to the database: they can import their collections of browser bookmarks. The system then automatically creates tags from the directory names in which the imported URLs were formerly stored in the browser. The authors explain how they dealt with the analysis of imported bookmarks and their related tags in the following paragraphs of this section.

[6] For example, *Bobdobr.ru*'s traffic rank on *alexa.com* is 471 (accessed on 3/12/2009).

Private bookmarks and their tags were, therefore, removed from the dataset, as their analysis would not contribute to the research questions. After their removal, a total of 110,740 stored URLs remained.

The authors also had to consider how to deal with imported bookmarks that were automatically tagged by the system (option 3). The automatic tagging process works as follows. If a URL has not been stored in a specific browser folder, the bookmark receives a single tag when imported to *BobrDobr.ru*, indicating that the URL belongs to "no category". If a URL has been stored in a specific browser folder, the directory and sub-directory names are assigned as tags. Since Veres [25] and Golder and Huberman [13] showed that hierarchical and taxonomical descriptions for URLs – equivalent to those that result from the automated tag creation taking the directory names of browser bookmarks as a basis – correspond to common tag choices, and as the authors found no indicators for assuming that these tags have a lower probability to be imitated, the authors decided to keep those bookmarks and tags in the dataset. Since URLs that were *solely* tagged automatically by the import mechanism of the system do not allow any statements about imitation of tags – that is, since no tags were created manually for these users by any user inside the system – the authors removed all URLs from their further analyses that solely received imported bookmarks. Furthermore, all "no category" tags were removed, because they were never assigned by hand – and, accordingly, never imitated – in the dataset and only appeared in imported bookmarks. In total, the authors removed 32,739 URLs; the resulting dataset contained 78,001 URLs and 299,786 tags.

In the next step, the authors selected all URLs that had received both types of bookmarks, "internal" and "external" ones. Some 661 URLs (.85%) met this requirement. The majority of URLs (77,282, or 99.08%) had received only "external" bookmarks. A small number of 58 URLs (.07%) had received no other bookmarks except for "internal" ones. Bookmarks that were created only internally or only externally were excluded from further analysis, because they would have distorted the results due to the fact that they are not two randomly assembled groups but rather had the risk of a self-selection bias by containing specific types of URLs linked to specific tagging behaviour. The composition of these groups and the resulting differences in tag sets would have always mixed with the effects of the bookmarking method. The authors, therefore, confined themselves to analyzing the 661 URLs (and their associated 5,799 unique tags) that provided clear insights into these effects.[7]

Table 1 depicts the distribution of bookmarks and tags for these 661 URLs. Each URL is represented in both columns, in the internal bookmarks as well as in the external bookmarks, as the URLs had received both types of bookmarks. However, a URL could have received 20 internal bookmarks, for example, but 30 external bookmarks and was therefore to be found in a different row for each column. A high proportion of URLs had only one or very few bookmarks. This is typical of the power law distribution of bookmarks over URLs [e.g. 13]. The table also shows that many more external bookmarks than internal ones were created.

[7] Although the number of 661 URLs might seem low in comparison to the total of 110,740 URLs in the original dataset, the sample size is still sufficiently large enough to draw meaningful statistical inferences about imitation behaviour and, hence, do not limit the analysis. Rather, this was a necessary step to ensure high data quality.

Table 1. Distribution of bookmarks and tags within the URLs[8]

	Internal bookmarks (copy method)		External bookmarks (browser button method)	
No. *m* of received bookmarks for a URL	No. of URLs with *m* bookmarks	No. of tags that had been assigned to these bookmarks (through copy method)	No. of URLs with *m* bookmarks	No. of tags that had been assigned to these bookmarks (through browser button method)
1	492	1320	485	2009
2-10	75	367	146	1203
11-20	46	405	3	56
21-30	48	505	3	102
31-40	0	0	2	115
41-50	0	0	19	979
51-60	0	0	3	184
Total	661	2597	661	4648

4.2 Methodology and Operationalization

The proposed hypotheses were tested using bivariate correlation analysis and the non-parametric Wilcoxon signed rank test. The variables used for these analyses were operationalized as follows.

For the test of ***H1***, the authors had to correlate the temporal order of tag assignment and the tag assignment frequencies of a URL.[9]

Concerning the operationalization of *overall tag frequency*, the authors had to consider that the absolute tag frequencies of a certain tag could not be compared between different URLs, because each had been assigned a different number of tags and a different number of unique tags. Ignoring the different group means could have led to

[8] Table 1 refers to the remaining dataset with both types of bookmarks (after eliminating private bookmarks and bookmarks for URLs that had received only imported bookmarks), examined in separate groups. One URL with 32 bookmarks was moved to the group with 21-30 bookmarks, as there was only one with more than 30 internal bookmarks.

[9] The bivariate distribution of the variables showed an unknown, non-linear relationship that differed between URLs. Other authors who have discussed social bookmarking with a focus on statistics [e.g., 28] have proposed transforming tag variables using the natural logarithm; this recommendation is based on many observations that tag distributions typically follow a power law / scale free distribution. The result would be linear relationships between the transformed variables and the applicability of linear models. For most of the URLs in the dataset analyzed, though, a logarithmic transformation was not viable, because no consistent scale free distribution could be observed for the URLs. This problem was then bypassed through the eventually used non-parametric transformation of the variables.

false conclusions regarding the overall relationship, on the aggregated level, of all URLs [26]. Thus, the absolute "internal" and "external" tag frequencies were rescaled as follows (for a detailed rationale for rescaling, such as "avoiding regression to the mean", see, e.g., [27]). First, the frequencies were transformed into ranks. The tag most often assigned to a URL received rank "1". When tags had been assigned in equal numbers, the resulting rank numbers were averaged for tied observations. Afterwards, percentile ranks were calculated using the formula $(r-1/2)/w$, where w is the number of observed tags for the underlying URL, and r denominates the rank of the value, with values from 1 to w. Finally, to obtain normal scores, the z-scores of the percentile ranks were computed, which result from standardizing the percentile ranks with respect to the mean and standard deviation of all the tags assigned to the underlying URL.

The *point in time* in which a tag t has been assigned to a specific URL u for the first time was measured as date in minutes. These timestamps were rank-transformed and standardized just as described for the frequencies, with the first assigned tag getting rank "1". This way, both of the rank orders were relative to the remaining ranks within the same URL and were then analyzed as two variables of a bivariate distribution.[10]

For the test of ***H2***, a variable was needed that would represent the difference of the internal and external assignment frequency of each tag for each URL. This variable was calculated by subtracting the absolute external tag frequency from the internal tag frequency. The differences were then rank transformed. High ranks were given to tags that were assigned internally much more often in absolute numbers than externally.

For the tests of ***H3*** and ***H4***, the variables t_{int} and t_{ext} indicate the average amount of tags assigned to a specific URL per bookmark, and dt_{int} and dt_{ext} indicate the average amount of unique tags for that same URL per bookmark; the index always shows whether tags assigned internally or externally are considered.

5 Results

Hypothesis ***H1*** postulates a positive correlation between the rank of temporal order in which a tag t has been assigned to a specific URL u for the first time and how many times t has been assigned to u overall in respect to all other tags of that URL.

For the hypothesis test, the authors first analyzed the "internal bookmarks". In a first step, the dataset was split into the four groups highlighted in Table 1 (i.e., one group for URLs with one bookmark, one group for URLs with 2-10 bookmarks, and so on). This decision was taken because the proportions of the tags frequencies of a URL shift as the URL gets more and more bookmarks (compare [13]). For example, the postulated effect of H1 cannot occur at all if a URL had received only a single bookmark. Therefore, the pooled analysis of a URL with one or two bookmarks with a URL with 40 bookmarks could not give meaningful insights, because the emerging

[10] Due to space limitations, the univariate distributions of the internal tag frequencies are not discussed here in detail. The univariate distribution (also for various groups as highlighted in the next section) showed a high concentration around the normal rank "0", similar to the distribution of the temporal ranks. This is plausible when considering that a large number of URLs had received only one bookmark with n tags, so that each of the tags had a total internal tag frequency of "1" and a normal rank of tag frequency (internal method) of "0".

descriptions for each URL were in different "stages of development". To reduce the number of groups, URLs with similar amounts of bookmarks were combined, as shown in Table 1.

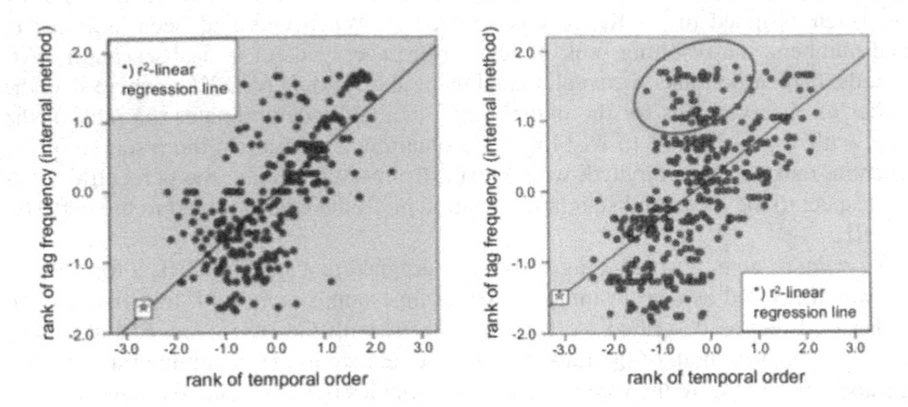

Fig. 2. Bivariate distribution "normalized rank of temporal order – normalized rank of tag frequency" for groups of internal tags. From left to right: Group with 11-20 and group with 21-30 bookmarks.

In line with the authors' expectations, for the group with one bookmark only, there was no statistically significant correlation between the normalized rank of tag frequency and the normalized rank of temporal order ($r(1320)=.02$; $p=.50$). A broader spectrum of tag frequencies was observed for the group with 2-10 bookmarks, showing, as expected a slightly positive correlation between temporal rank order and tag frequency ($r(367)=.24$; $p=.00$), but also a broad dispersion with no obvious linearity. For the group with 11-20 bookmarks, the resulting correlation came closer to the expected linear relationship (see the left graph of Figure 2) with a relatively high coefficient ($r(405)=.71$; $p=.00$). This group also contained a number of outliers, tags that had been assigned seldom but had been assigned relatively early to the observed URL. The results of the group with 21-30 bookmarks were similar ($r(505)=.59$; $p=.00$), only that outliers occurred more often, and in another form as well: there were some tags that were assigned relatively late but were still quite common; the oval in the second graph of Figure 2 highlights them. In summary, ***H1*** is supported for the internal bookmarks.

In a second step, the authors tested ***H1***, analyzing the external bookmarks for the groups from Table 1 (see Figure 3).

As for the internal bookmarks, the result for group 1 (URLs with only one bookmark) of the external bookmarks did not indicate a correlation between the rank of temporal order in which a tag t has been assigned to a specific URL u for the first time and how many times t has been assigned to u overall ($r(2009)=$ -.00; $p=.84$). The distribution of the group with 2-10 bookmarks showed no linear relationship (see Figure 3) although Pearson's r was significant ($r(1203)=.32$; p=.00). Many tags received high frequency ranks (i.e., they were rarely assigned), independently from the point in time they had been assigned. A considerable amount of tags had only been

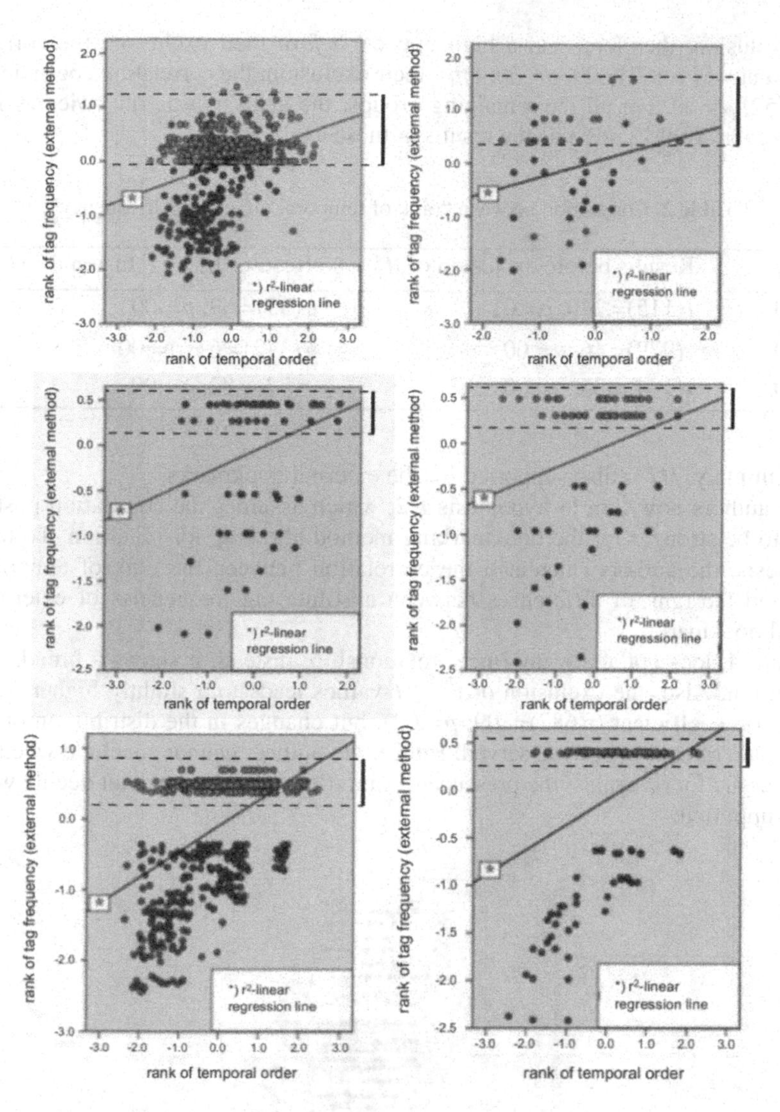

Fig. 3. Bivariate distribution "rank of temporal order – rank of tag frequency (external method)" for the groups of external tags. From left to right, from above to below: Group with a) 2-10, b) 11-20, c) 21-30, d) 31-40, e) 41-50, f) 51-60 bookmarks.

assigned once (marked by the dotted-line brackets in the figures). They may be very individual tags, so specific that no other user would associate them with the underlying URL, independently from the point in time they were assigned. We call these values "*tf1*-values" for "tag frequency 1". When excluding these values from the analysis, the correlation coefficient rose to $r(403)=.39$, $p=.00$, which could be an indicator for the appropriateness of the authors' interpretation. For the group with 1120 bookmarks, the correlation was positive ($r(56)=.29$; $p=.03$) and the exclusion of the *tf1*-values did not lead to a higher coefficient. In group 21-30, *tf1*-values were highly represented and their exclusion, therefore, had a high impact: *before* their exclusion, the correlation coefficient was $r(102)=.29$, $p=.00$; *after* their exclusion, the correlation coefficient was $r(33)=.52$, $p=.00$. For all the remaining groups, the analysis was run twice as for the other groups. Table 2 presents the results of these analyses.

Table 2. Correlation between "rank of temporal order" and "frequency"

Group	Results before exclusion of *tf1*	Results after exclusion of *tf1*
31-40	$r(115)=.240$, $p=.01$	$r(35)=.59$, $p=.00$
41-50	$r(979)=.46$, $p=.00$	$r(372)=.61$, $p=.00$
51-60	$r(184)=.35$, $p=.00$	$r(57)=.82$, $p=.00$

In summary, ***H1*** is also supported for the external bookmarks.

The authors now turn to hypothesis ***H2***, which assumes the correlation postulated in ***H1*** to be stronger for the bookmarking method allowing for imitation. To test this hypothesis, the authors calculated the correlation between the rank of temporal tag order and the rank of differences between absolute tag frequencies of external and internal bookmarks.

Figure 4 does not show any linear relationship; instead, it shows a broad, regular distribution. Also, the exclusion of the *tf1*-values leads to a slightly higher negative correlation coefficient ($r(687)=-.09$; $p=.02$), but changes in the distribution or a particular pattern could not be observed. Hence, the authors cannot conclude whether the results argue for or against the presence of imitation and hence cannot decide whether ***H2*** is supported.

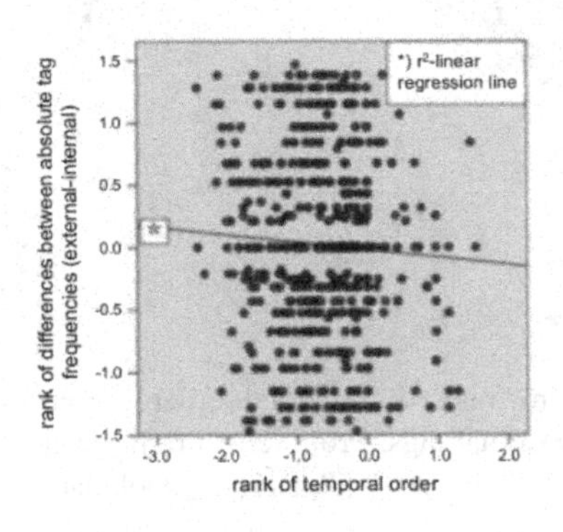

Fig. 4. Bivariate distribution "rank of temporal order – rank of differences between absolute tag frequencies (external-internal)"

H3 postulates that the average amount of tags a user assigns to a URL increases when the system allows that user to see other users' tags, while ***H4*** postulates that the average amount of *different* tags assigned to a URL by all users decreases.

A Wilcoxon signed rank test was conducted to compare the 661 URLs that received both the internal and external types of bookmarks. For this test, the authors calculated the differences between the values "number of tags" and "number of different tags" resulting from the internal and the external method for each URL. The Wilcoxon signed rank test showed that fewer tags were assigned through the internal than through the external bookmarking method (z=-7.80; p=.00), that is, more URLs received more tags from the external method than the other way around. Hence, when imitation was possible, users did not assign more tags to a URL, even if they may have had more choices. ***H3***, therefore, is not supported. In contrast, the test results of the second Wilcoxon signed rank test (z=-13.54; p=.00) point favourably to hypothesis ***H4***, which assumes that the variation of assigned tags decreases when imitation is possible. More URLs received less diversified tags from the internal method than the other way around. ***H4***, therefore, is supported. Of course, the lower tag variation for the internal method is due in large part to the smaller number of tags assigned. Note, however, that the negative value of z is considerably larger for the different tags assigned, which speaks to a second cause, namely imitation.

6 Summary and Conclusions

The aim of this study was to analyze the impact of a feature allowing for imitation provided by a social bookmarking system on its resulting metadata. The authors were able to show that tags that were assigned early were assigned more often, coming closer to a linear relationship the more bookmarks were assigned to a URL. This was true for internally as well as for externally assigned tags (***H1***). The authors were not able to show a clear increase in the correlation of the time of assignment and the frequency of a tag when the imitation feature was present, meaning that popular tags get more popular when imitation is possible (***H2***). Nonetheless, comparing the bivariate distributions in the internal and external case, it became clear that with the internal method tags that were assigned only once were largely lacking. This is an indicator that users neglect some of their personal, very individual tags when tag recommendations are available.

The authors' assumption that users assign more tags when provided with other users' tag choices clearly had to be rejected (***H3***), since the results indicated the opposite. One reason for this may be that users see little or no need to think about tags of their own when presented with a choice of popular tags from which to choose conveniently. This would be in line with the principle of least cognitive effort described by Munk et al. [4]. Consequently, users adopt only the suggested set of tags. This explanation fits with the findings that, when imitation is possible, the number of different tags assigned to a URL decreases partly due to the drop in the number of tags assigned overall (***H4***). The other element of the decrease of different tags could then be explained by users constraining themselves to the sole use of suggested tags.

Overall, these findings point to strong effects of imitation on the emerging folksonomies, leading to a less pluralistic collective description of content. This might be interpreted as an overrepresentation of early tags leading to the elimination of individual information and thus a decreasing quality of the folksonomy. But, as this effect could not be shown unequivocally, the less different keywords can also mean a unification of synonyms or different spelling. In such a case, users who used different tags for the description of exactly the same concept would align their tags with the "standard" for that concept set in the most popular tags. This could be seen as a positive effect of imitation.

Still, these findings have important implications for researchers and practitioners who analyze or intend to use folksonomies for content classification. The feature of the system analyzed in this study, for example, aimed only at increasing convenience for users during the tagging process. However, this feature also had a large impact on the imitation of tags, and hence also on the variety of tags and the resulting folksonomy as a whole. Therefore, the authors recommend that system designers always consider how features they implement may affect the quality of the resulting folksonomies, particularly when they would like to utilize full cognitive input of all users.

The simplest solution to ensure a higher quality of collective tagging would be the general omission of tag suggestions based on aggregated taggings. This, however, is not viable in systems that face tough competition and that rely on simple, effortless usage to attract users. Another option would be a "type-ahead" feature that requires users to begin typing two to three letters and then offers suggestions for completion from the pool of most popular tags. In this way, users would have to tap into their own cognitive models and would not be guided in one direction or the other before they think themselves about possible tags. To benefit from positive imitation effects, the suggestion for a tag should be the most popular synonym and spelling form. More elaborate mechanisms produce tag suggestions based on the content of bookmarked websites or files; others draw on the tri-partite links between resources, tags, and users (i.e, graph-based tag suggestion). These solutions rely on the most popular tags only to a small part or not at all and are therefore not likely to produce information cascades.

In addition to this managerial insight, this study provides some insights for future scientific work on the subject. The authors are not aware of any study that tests statistically any formal hypotheses of imitation of tags in social bookmarking systems with empirical field data.

As with any empiricial study, this work is subject to limitations. The authors do not consider these limitations to void any results so long as the reader remains aware of them when interpreting the results. In fact, they suggest either some future research that examines collective intelligence in social bookmarking systems or provide additional insights about user behavior in social bookmarking systems.

First, indicators speak to the presence of spam in the raw data (cf. also [13]); for example, a large number of tags were never copied with respect to one URL but assigned to a great number of different URLs. No method to avoid spam completely in social bookmarking systems exists, so only an experimental design or a system that can guarantee a human-only userbase could provide a folksonomy created completely absent the influence of spam.

Second, the small number of internally created bookmarks and the resulting small number of URLs with both types of bookmarks (661 URLs; 0.85%) was surprising. The majority of URLs (77,282 URLs, i.e. 99.08%) had received only "external" bookmarks. The low affinity to the copying of other users' bookmarks could be explained by the users' preference to inspect website content themselves before tagging it, and therefore having a preference for the external bookmarking method. It seems that far fewer users than expected browse the system to "stumble upon" bookmarks that might interest them, but use the external bookmarking button only to save addresses they found for themselves on the web.

Third, considering these findings, it is also legitimate to raise objections to the implicit assumption that the two bookmarking methods do not just differ only with respect to suggesting or not suggesting tags. For example, one could argue that some people browse for bookmarks internally and then click on a link to inspect the website, after which they bookmark the website using the external method. Still, in this case, the user would not see the most popular tags either, as these are only displayed when bookmarking internally. One might also argue that these users would remember those tags they saw beforehand and assign them as well when bookmarking URLs with the external method. Both threats to reliability would lead only to a higher than postulated similarity rate between externally and internally assigned tags. This would reduce the power of the tests employed, and hence it would be more difficult to find support for the hypothesized effects of imitation. However, if significant effects of imitation are found (as in this study), the reader can have confidence in these findings, because the reliability threat does not threaten validity when a difference is found.[11]

Fourth, it should be mentioned that the cultural background of the userbase of *BobrDobr.ru* could have influenced the results. It cannot be precluded that a different affinity for imitation is present in different cultural contexts.[12] Still, as none of the literature on information cascades or rational imitation reviewed by the authors offers any hints as to cultural influences on the postulated effects, we assume that the basic mechanism of imitative behavior is universal in such systems even if the extent of imitation can be different.

Fifth, one might argue that users are heterogeneous regarding their preferred bookmarking method. In this case, imitation behavior linked to the internal bookmarking method could be an effect of the user type preferring this method and their affinity to imitation rather than an effect of the bookmarking method itself. Future research should analyze whether preferences for a certain bookmarking method are linked to a certain type of user. The best way to factor out such effects unambiguously would be an experimental design with randomly assembled groups of users, one bookmarking with and one without suggestions. The authors suppose that such a lab experiment would provide additional confidence in the findings from this field experiment.

The authors hope that this research will assist other researchers in conducting these types of studies and form the basis for substantial future research into imitation of tags and information cascades in social bookmarking services. Further, the authors hope that this research provides useful insights for managers, librarians, and other practitioners who use folksonomies for content classification and who must design social bookmarking systems that will lead to high-quality folksonomies.

References

1. Marlow, C., Naaman, M., Boyd, D., Davis, M.: HT06, Tagging Paper, Taxonomy, Flickr, Academic Article, ToRead. In: 17th conference on Hypertext and Hypermedia, Odense, Denmark, pp. 31–40. Association for Computing Machinery, New York (2006)

[11] The real impact of imitation should be rather stronger.

[12] One could argue, for example, that the states of the former U.S.S.R are more prone to collectivism, which, in turn, favors imitative behavior [compare e.g. 29].

2. Vander Wal, T.: Folksonomy Definition and Wikipedia. Blog entry (November 2005), http://www.vanderwal.net/random/entrysel.php?blog=1750 (accessed on September 9, 2009)
3. Vander Wal, T.: Folksonomy: Coinage and Definition. Blog entry (February 2007), http://vanderwal.net/folksonomy.html (accessed on September 9, 2007)
4. Munk, T.B., Mørk, K.: Folksonomy, the Power Law & the Significance of the Least Effort. Knowledge Organization 34(1), 16–33 (2007)
5. Al-Khalifa, H.S., Davis, H.C.: Replacing the Monolithic LOM: A Folksonomic Approach. In: 7th IEEE International Conference on Advanced Learning Technologies (ICALT 2007), Niigata, Japan (2007)
6. Tsui, E., Wang, W.M., Cheung, C.F., Lau, A.S.M.: A Concept-Relationship Acquisition and Inference Approach for Hierarchical Taxonomy Construction From Tags. Information Processing & Management 46(1) (in print, 2010)
7. Van Damme, C., Coenen, T., Vandijck, E.: Turning a Corporate Folksonomy into a Lightweight Corporate Ontology. In: Abramowicz, F. (ed.) BIS 2008. Lecture Notes in Business Information Processing, vol. 7, pp. 36–47 (2008)
8. Farooq, U., Song, Y., Carroll, J.M., Giles, C.L.: Social Bookmarking for Scholarly Digital Libraries. IEEE Internet Computing 11(6), 29–35 (2007)
9. Li, X.R., Snoek, C.G.M., Worring, M.: Learning Social Tag Relevance by Neighbor Voting. IEEE Transactions on Multimedia 11(7), 1310–1322 (2009)
10. Goshal, G., Zlatic, V., Caldarelli, G., Newman, M.E.J.: Random hypergraphs and their applications. Physical Review E 79(6) (2009) (in print)
11. Zhang, Z.K., Zhou, T., Zhang, Y.C.: Personalized Recommendation via Integrated Diffusion on User-Item-Tag Tripartite Graphs. Physica A - Statistical Mechanics and Its Applications 389(1), 179–186 (2010)
12. Boast, R., Bravo, M., Srinivasan, R.: Return to Babel: Emergent Diversity, Digital Resources, and Local Knowledge. Information Society 23(5), 395–403 (2007)
13. Golder, S.A., Huberman, B.A.: Usage Patterns of Collaborative Tagging Systems. Journal of Information Science 32(2), 198–208 (2006)
14. Heymann, P., Koutrika, G., Garcia-Molina, H.: Can Social Bookmarking Improve Web Search? In: International Conference on Web Search and Web Data Mining (WSDM 2008), Palo Alto, California, USA, pp. 195–206 (2008)
15. Bao, S., Wu, X., Fei, B., Xue, G., Su, Z., Yu, Y.: Optimizing Web Search Using Social Annotations. In: 16th International Conference on World Wide Web (WWW 2007), Banff, Alberta, Canada, May 8-12 (2007)
16. Macgregor, G., McCulloch, E.: Collaborative Tagging as a Knowledge Organisation and Resource Discovery Tool. Library Review 55(5), 291–300 (2006)
17. Specia, L., Motta, E.: Integrating Folksonomies with the Semantic Web. In: Franconi, E., Kifer, M., May, W. (eds.) ESWC 2007. LNCS, vol. 4519, pp. 624–639. Springer, Heidelberg (2007)
18. Hotho, A., Jäschke, R., Schmitz, C., Stumme, G.: Information Retrieval in Folksonomies: Search and Ranking. In: Sure, Y., Domingue, J. (eds.) ESWC 2006. LNCS, vol. 4011, pp. 411–426. Springer, Heidelberg (2006)
19. Arakji, R., Benbunan-Fich, R., Koufaris, M.: Exploring Contributions of Public Resources in Social Bookmarking Systems. Decision Support Systems 47(3), 245–253 (2009)
20. Rosch, E., Mervis, C.B., Gray, W., Johnson, D., Boyes-Braem, P.: Basic Objects in Natural Categories. Cognitive Psychology 8(3), 382–439 (1976)
21. Tanaka, J.W., Taylor, M.: Object Categories and Expertise: Is the Basic Level in the Eye of the Beholder? Cognitive Psychology 23(3), 457–482 (1991)

22. Cialdini, R.B.: Influence: Science and Practice. Harper Collins College Publishers, New York (1993)
23. Hedström, P.: Rational Imitation. In: Social Mechanisms: An Analytic Approach to Social Theory. Cambridge University Press, Cambridge (1998)
24. Bikhchandani, S., Hirshleifer, D., Welch, I.: A Theory of Fads, Fashion, Custom, and Cultural Change as Informational Cascades. The Journal of Political Economy 100(5), 992–1026 (1992)
25. Veres, C.: The Language of Folksonomies: What Tags Reveal About User Classification. In: Kop, C., Fliedl, G., Mayr, H.C., Métais, E. (eds.) NLDB 2006. LNCS, vol. 3999, pp. 58–69. Springer, Heidelberg (2006)
26. Boyd, L.H., Iversen, G.R.: Contextual Analysis: Concepts and Statistical Techniques, Wadsworth, Belmont, California, USA (1979)
27. Jones, Q., Ravid, G., Rafaeli, S.: Information Overload and the Message Dynamics of Online Interaction Spaces: A Theoretical Model and Empirical Exploration. Information Systems Research 15(2), 194–210 (2004)
28. Raban, D.R., Rabin, E.: Statistical Inference from Power Law Distributed Web-Based Social Interactions. Internet Research 19(3), 266–278 (2009)
29. Marron, D.B., Steel, D.G.: Which Countries Protect Intellectual Property? The Case of Software Piracy. Economic Inquiry 38(2), 159–174 (2000)

Measuring and Analyzing the Openness of the Web2.0 Service Network for Improving the Innovation Capacity of the Web2.0 System through Collective Intelligence

Kibae Kim, Jörn Altmann, and Junseok Hwang

Technology Management, Economics and Policy Program
Department of Industrial Engineering
College of Engineering, Seoul National University
599 Gwanak-Ro, Gwanak-Gu, Seoul 151-744, South-Korea
kibaejjang@gmail.com, jorn.altmann@acm.org, junhwang@snu.ac.kr

Abstract. Web2.0 users can create new services by combining existing Web2.0 services that offer open programming interfaces. This system of service composition forms a network, which we call the Web2.0 service network. A node of the Web2.0 service network represents a service. A link between two nodes exists, if another Web2.0 service (i.e. mashup) uses the linked services. The Web2.0 service network can be understood as an innovation system that creates value through the composition of services, representing the collective intelligence of users. Within this paper, we analyze the openness of the Web2.0 service network. Openness, which is an indicator for the innovation potential of a network, is measured using the Enhanced-EIS-Indexes. These indexes are based on Krackhardt and Stern's EI-Index. The analysis results of the indexes show that the Web2.0 service network is not as open as the evolutionary analysis of the Web2.0 service network suggested. The slight closeness of the Web2.0 service network has been identified by the Agent Behavior Index EIS_a, which highlighted that relatively more links are created within subgroups than between subgroups. It indicates that factors such as service ownership and type of service have an impact on innovation within the network.

Keywords: Social network analysis, index, network science, subgroup structure, Web2.0 system, service composition, collective intelligence, Web2.0 service network, performance evaluation, innovation, empirical data analysis.

1 Introduction

In the Web2.0 system, users can combine Web2.0 services that offer open application programming interfaces (APIs). The composed Web2.0 services (which are called *mashups*) provide content, which can again be shared with other Internet users if it provides an open API as well. For making Web2.0 service creation simple, service providers such as Google and Yahoo provide platforms that enable users to create a variety of Web2.0 services based on their ideas and content [1] [21] [22].

The Web2.0 system evolves because of the interactions between the stakeholders in the system. These interactions of the agents organize the Web2.0 service network,

T.J. Bastiaens, U. Baumöl, and B.J. Krämer (Eds.): On Collective Intelligence, AISC 76, pp. 93–105.
springerlink.com

which consists of nodes representing Web2.0 services and links their use by other services. The analysis of the evolutionary trends of this network could identify the drivers behind it [3]. Initial work on this topic has been performed [4] [10]. The authors analyzed the network to understand the mechanism of its structural evolution.

However, since the Web2.0 service composition itself also represents value creation, further investigation is needed. In particular, it needs to be analyzed whether the collective intelligence can be fully utilized for Web2.0 service composition, since the Web2.0 system is considered a typical example of harnessing collective intelligence [1]. The Web2.0 system enhances the pool of knowledge by allowing users to participate, create ideas, and realize their ideas through simple passing of knowledge [19]. It allows users to be innovative.

A key enabler of collective intelligence (and, therefore, innovation) is openness. It is a prerequisite together with three other enablers, namely sharing, acting globally, and peering, in order to allow participants to exchange information freely [20].

The openness of Web2.0 services distinguishes it from the first generation Web [1], in which service providers only published their service offerings and, after a contractual agreement with a customer, provided a service [2]. However, no study of the Web2.0 system has addressed the aspect of openness of the Web2.0 system in detail. In particular, no research exists that investigates the importance of openness on the full utilization of collective intelligence of Web2.0 users.

The openness of the Web2.0 service network is important since collective intelligence (and consequently innovation) can only work through the exchange of resources. Even if technology promotes user interaction, social and economic barriers may prohibit the perfect openness between competitors. For example, if the Web2.0 system has limited openness, then Web2.0 innovation is achieved in separated groups (subgroups) only but not across subgroups. In this case, the Web2.0 system is nothing but a collection of partitioned innovation groups, limiting harnessing the collective intelligence of the Web2.0 users.

In order to address this issue, this paper analyzes the impact of subgroups on the actual openness of the Web2.0 service network. The Web2.0 service network defines Web2.0 services as nodes. A link between two nodes exists, if a mashup between those two Web2.0 services has been constructed. Subgroups, which are defined through criteria such as ownership of Web2.0 services (e.g. Google, Yahoo, and Amazon) and the type of service (e.g. shopping services, map services), classify Web2.0 services into different subgroups.

In particular, we introduce the Enhanced-EIS-Indexes to measure the openness of the Web2.0 service network. The three indexes are based on Krackhardt and Stern's EI-Index. The idea behind these indexes is to compare their values, which represent the ratios of links that are generated between subgroups (i.e. external relationships), within subgroups (i.e. internal relationships), and for sole nodes (i.e. self relationships). The ratios indicate how open a network is.

The following section provides background on the Web2.0 system as an innovation system, its interrelation with collective intelligence, and how social network analysis can help to analyze it. In Section 3, we introduce the Enhanced-EIS-Indexes. Based on these indexes, the Web2.0 service network is analyzed in Section 4. The paper concludes with an analysis of the results in Section 5, stating that the Web2.0 service network is not as open as suggested by the industry but rather slightly closed.

2 Innovation, Collective Intelligence, and Network Analysis

2.1 Innovation Capacity of the Web2.0 System through Collective Intelligence

As many examples have shown, high quality and high quantity of innovation emerges in an environment, in which agents can exchange resources beyond the boundaries of formal organizations [6] [7]. For example, in the Silicon Valley, innovation is high, since knowledge flows freely between R&D centers and start-up companies due to high mobility of human resources [5]. Chesbrough analyzed this in 2003 and defined innovation that is based on knowledge sharing as *Open Innovation* [6]. It differs from the old way of innovation, which achieves innovation through vertical integration of research, development, and marketing.

Companies that follow the open innovation approach do not open all their resources they have produced [7] [9]. For example, some open source developers for embedded Linux code reveal part of their code in order to attract potential customer. Code that provides more functionality is sold, helping them to be competitive [8]. In this system, the proprietor of core technology gains profit from an increased market size that cooperators utilizing its free resources have cultivated [7]. The proprietor, who shares its resources, benefits through collective intelligence. In information science, collective intelligence is defined as "the capacity of human collectives to engage in intellectual cooperation in order to create, innovate and invent" [23]. It requires opening access to a resource and sharing it with competitors [20].

Considering the Web2.0 System, the technology of the Web2.0 system promotes cooperation between competing firms on a single platform, enabling the use of collective intelligence of all users of the Web2.0 system and, therefore, fostering innovation. The innovation within the Web2.0 system (i.e. development of mashups) increases user utility and, if a mashup comes with open APIs, even benefits the development of further Web2.0 services. Therefore, mashups (i.e. relationships between Web2.0 services) of different firms represent innovation while the Web2.0 services can be independent and belong to different firms.

2.2 Value Chain of the Web2.0 System

The Web2.0 value chain has two types of stakeholders: the Web2.0 service provider and the user (Fig. 1). The service providers, which provide a platform for executing the Web2.0 services, can be further grouped into providers that offer Web2.0 services with open APIs and those that offer Web2.0 services without APIs (which is similar to the first generation Web). Both kinds of Web2.0 services are open to all users. While Web2.0 without an open API can only be consumed, Web2.0 services with open APIs allow users to create new services by combining the Web2.0 service with other Web2.0 services and their own content. If they do so, those users (*Commercial Developer* of Fig. 1) become practically Web2.0 service providers. If users only add their own content to an existing Web2.0 service, they are not considered service provider in this model (*Consumers Developing Web Sites* of Fig. 1).

The motivation of users to participate in this value creation system varies, ranging from enjoying it as a hobbyist to earning money as a commercial developer. For example, Web2.0 service providers that offer services with open APIs can generate revenues by sharing profit with commercial developers, who utilize their services.

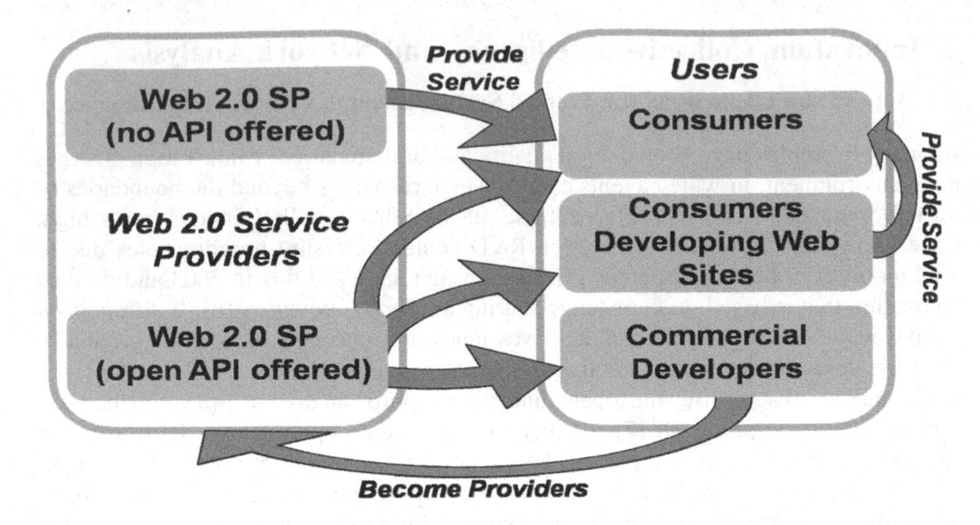

Fig. 1. Provider-user relationship model for Web2.0 services

2.3 Quality of Innovation through Openness

By allowing users to participate in the development of Web2.0 services, value is created within the Web2.0 system. The innovation comes from combining Web2.0 services in new ways. Assuming that the combination of Web2.0 services can be rated according to the number of Web2.0 services per subgroup and the type of Web2.0 services (i.e. subgroup), the quality of innovation can be illustrated in an $2n$-dimensional space, where n is the number of subgroups. That means each 2-dimensional plane represents one subgroup. Fig. 2 depicts an example in the 3-dimensional space.

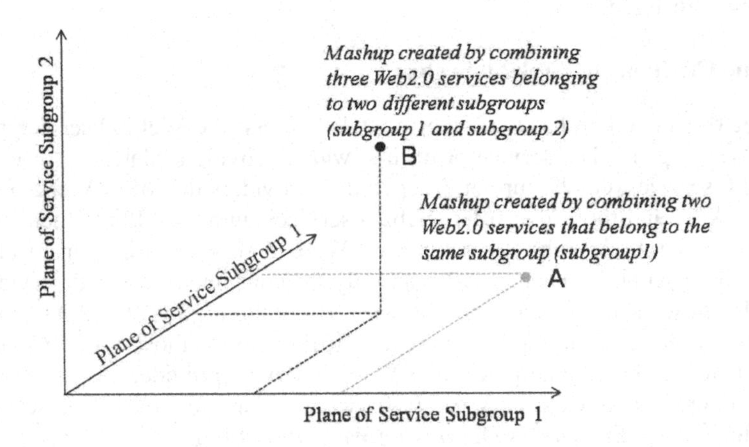

Fig. 2. Example of two mashups located in two service subgroup planes

Fig. 2 shows a mashup *A*, which has been developed by combining two Web2.0 services of the same subgroup (i.e. subgroup 1). The mashup is located in one plane. Fig. 2 also depicts another mashup *B*, which has been developed with three Web2.0 services belonging to two different types of services (i.e. subgroup 1 and subgroup 2). The mashup *B* can be considered more innovative than mashup *A*, since it requires openness between subgroup 1 and subgroup 2.

2.4 Social Network Analysis of the Web2.0 Service Network

After we have described how innovation is created and innovation flows are established, we need to investigate how innovation of the Web2.0 system can be measured. We need to analyze whether subgroups are open or not, i.e. whether the industry allows innovation to happen freely across different Web2.0 service subgroups (e.g. the ownership of Web2.0 services).

Essential innovation activities are not only the result of the creation of resources but also the result of the combination of resources [6]. For analyzing the resource combinations (e.g. resource sharing between firms), a network approach is useful. This approach measures the position and role of agents in a social network as well as the characteristics of the network structure [11] [12]. An agent's position within a social structure and the structural characteristics of the network can be regarded as social capital, since they determine the interactions between agents [13] [14]. For example, an agent with many social relationships can influence other agents and disseminate information quickly.

To distinguish different characteristics of social capital, Walter et al. (2007) surveyed prior research on structural indicators (e.g. network density, structural hole, and centrality) with respect to inter-firm and intra-firm innovation [15]. They argued that a hub-client structure with a small number of core firms as hubs is most efficient for inter-firm innovation activities while a structure with many redundant ties between departments (e.g., subgroups) is an effective structure for intra-firm innovation activities.

To investigate the effect of openness of subgroups on the performance of an entire organization, Krackhardt and Stern (1988) developed the EI-Index [16]. The EI-Index is defined as the difference between the number of external and internal links divided by the total number of links. An external link is defined as a link between subgroups while an internal link is defined as a link within a subgroup. The index is *1*, if all members only have relationships between their subgroups, and *-1* if the relationships exist only within subgroups. Krackhardt and Stern applied this index to friendship networks of organizations [16]. The result showed that organizations, in which subgroup members interacted with members of other subgroups, yielded a high performance in a crisis. For analyzing the effect of interaction across boundaries of subgroups on the innovation of organizations, the EI-Index has also been applied to the communication network of the Knowledge Management Group at Fraunhofer-Gesellschaft [17] and to the research collaboration networks of multi-national R&D centers in Scandinavia [18].

3 An Index for Measuring Openness

3.1 The Web2.0 Service Network

The Web2.0 service network is defined as a set of nodes, representing Web2.0 services with open APIs, and links indicating the existence of a mashup that uses the nodes being connected [4]. In addition to belonging to the Web2.0 service network, Web2.0 services are also classified into groups (subgroups). The classification criterion allocates each Web2.0 service into exactly one subgroup.

Within this paper, we consider two subgroup classification criteria. First, Web2.0 services are classified according to their *type of service* (e.g. subgroup *Map Services*, subgroup *Picture Storage Services*, and subgroup *Shopping Services*). For example, the Web2.0 services Google Maps and Yahoo Maps belong to the subgroup *Map Services*, and Google Checkout and Yahoo Shopping are grouped into the subgroup *Shopping Services*.

Second, Web2.0 services are grouped according to their *ownership*, i.e. the company that owns the Web2.0 service. With respect to this classification, the Web2.0 services Google Maps and Google Checkout are classified into the subgroup *Google*, and Yahoo Search and Yahoo Shopping are placed in the subgroup *Yahoo*.

3.2 Characteristics of the Links of the Web2.0 Service Network

Our model distinguishes the types of relationships between Web2.0 services of a Web2.0 service network. The types are self-relationships, internal relationships, and external relationships (Fig. 3). A Web2.0 service has a self-relationship, if a mashup exists that has been developed based on only this Web2.0 service. A link between two Web2.0 services represents an internal relationship, if both Web2.0 services, which have been used by a mashup, belong to the same subgroup. A relationship between two Web2.0 services is considered to be an external relationship, if these two Web2.0 services belong to different subgroups. In Fig. 3, the italicized numbers next to the links (ties) represent weights of the ties. For example, the value of 3 next to the external relationship means that the two nodes are connected through three mashups.

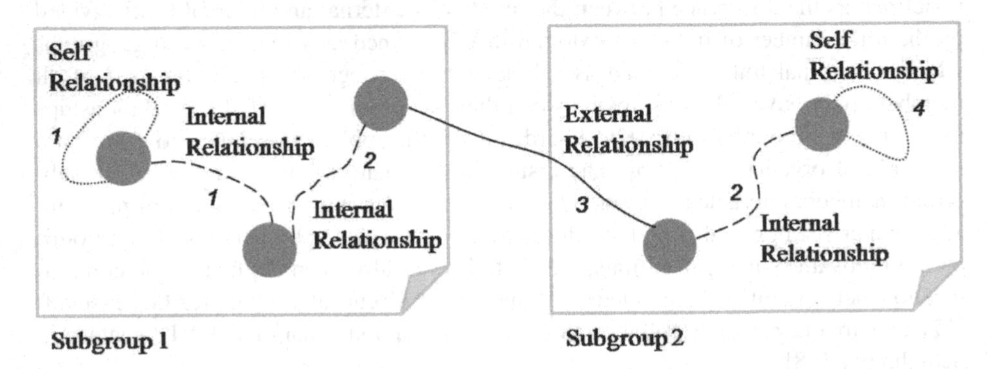

Fig. 3. Example of self relationships, internal relationships, and external relationships in a social network such as the Web2.0 service network

3.3 Measures for Evaluating the Openness of Networks

In order to analyze the Web2.0 service network, we use two types of measures. The first type of measure is used to compare the relative relationships of the three types of links (i.e. self relationship, internal relationship, external relationship). The second type of measure, which is an enhancement of the Krackhardt and Stern Index, is used to investigate the ratio between external links and the internal and self links (Section 3.4).

3.4 Relative Relationship Value

The first type of measure, which is called relative relationship value, helps weighting the relative importance of the three types of links. It ranges between 0 and 1. The relative relationship value of the self (or internal or external) links is defined as the ratio of the number of self (or internal or external) relationships and the total number of links in the network.

In order to explain the structural characteristic of the network through a comparison of these relative relationship values, we also introduce the two terms *dominant* and *superior*. A relative relationship value is called *superior*, if it is larger than another relative relationship value. A relative relationship value is called *dominant*, if it is larger than the sum of the other two values. That is, a superior (or dominant) relative relationship value indicates that its characteristic is strongly (or very strongly) represented in the network.

3.5 The Enhanced-EIS-Indexes

Applying Krackhardt and Stern's EI-Index to the Web2.0 service network has two limitations. First, binary link weights, as used by Krackhardt and Stern's EI-Index, only show the existence of relationships but do not represent the strength of ties between nodes. Therefore, the same combination of Web2.0 services used for developing different mashups would not be considered. However, since the strengths of links express the importance of nodes, this information is valuable for the analysis of the Web2.0 service network.

Second, the EI-Index, which identifies only internal and external links, does not allow representing mashups that have been developed based on one single Web2.0 service. These mashups have been created by simply adding information to one existing Web2.0 service. For the analysis of the Web2.0 service network, self-relationships represent a significant fraction and, therefore, have to be considered in the analysis of the network.

The Web2.0 service network can be analyzed comprehensively, if we enhance Krackhardt and Stern's EI-Index by including self-relationship and weighted links. Following the format of Krackhardt and Stern's EI-Index, we can define the Enhanced-EIS-Index EIS_r as:

$$EIS_r = (E - I - S) / (E + I + S)\,, \qquad (1)$$

where, E, I, and S are the number of external, internal, and self-relationships of the network, respectively.

To distinguish the effects of the social network structure on the Enhanced-EIS-Index EIS_r, we introduce two more indices, which are similar to the Enhanced-EIS-Index. The first index considers the network size (i.e. the number of nodes) and the subgroup structure by determining the maximum possible number of dichotomous external (internal, and self) relationships that a network can generate at most. For example, the network in Fig. 3 has 5 nodes, of which 3 nodes belong to subgroup 1 and 2 nodes to subgroup 2. Therefore, the network can at most generate 6 external dichotomous relationships, 4 internal dichotomous relationships, and 5 self dichotomous relationships. By using the maximum possible number of external E*, internal I*, and self S* links of the dichotomous network [16], we define the *Enhanced-EIS$_s$-Index* as:

$$EIS_s = (E^* - I^* - S^*) / (E^* + I^* + S^*) . \quad (2)$$

The resulting EIS_s value indicates the openness of a network that the subgroup structure of the network allows. Since the maximum possible number of links depends on the distribution of the size of the subgroups, we also call this index *Subgroup Structure Index*. Considering our example, we get EIS_s = (6-4-5) / (6+4+5) = - 0.2.

The second index considers the normalized numbers of relationships, while excluding the effect of the subgroup structure. The normalized number of external (internal, and self) relationships e (i, and s) is defined as ratio of the number of external (internal, and self) relationships and the maximum possible number of external (internal, and self) relationships. For example, the network shown in Fig. 3 has 1 external relationship, 3 internal relationships, and 2 self relationships. And it can hold at most 6 external relationships, 4 internal relationships, and 5 self relationships. Hence, the normalized number of external (internal, and self) relationships is 0.166 (0.75, and 0.4). By applying the normalized numbers of external, internal and self links (e, i, and s), we can define the *Enhanced-EIS$_a$-Index* as:

$$EIS_a = (e - i - s) / (e + i + s) . \quad (3)$$

Since the effect of the subgroup structure is neutralized in this index, the resulting EIS_a value represents the openness of a network without the subgroup structure effect. Since the normalized numbers of links represent the average numbers of relationships of agents in the network, we also call this index *Agent Behavior Index*. In our example, the Index EIS_a = (0.166-0.75-0.4) / (0.166+0.75+0.4) = -0.7468.

4 Analysis

4.1 Description of Empirical Data

Data has been gathered from the Web site http://www.programmableweb.com, which lists Web2.0 services that have been registered by the owners of the services. In total, 2374 Web services had registered between September 1st, 2005 and May 31st, 2007, of which 445 are Web2.0 services that offer open APIs and 1929 are mashups. Besides, 8 mashups also provide open APIs. However, 226 Web2.0 services were eliminated from the data set, since they are not used in any mashup during the survey period. As a result, only 219 services with open APIs are used in the analysis.

The resulting Web2.0 service network can be classified into 143 subgroups with respect to the *ownership* criterion and into 47 subgroups with respect to the *type of service* criterion. The size of these two sets of subgroups is heterogeneously distributed. Among the 143 subgroups created through the ownership criterion, 134 companies provided only one Web2.0 service. Only nine companies (e.g. Google, Yahoo, and StrikeIron) provided more than two Web2.0 services (Fig. 4).

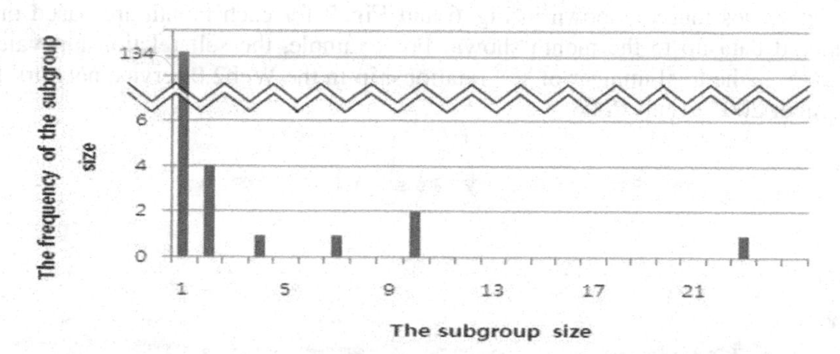

Fig. 4. Distribution of the subgroup size for subgroups based on the ownership criterion

The results that are shown in Fig. 4 indicate that the distribution of Web2.0 services is not uniform over all companies. That is, a small amount of subgroups have a majority of Web2.0 services with open APIs, and a large part of subgroups includes only few Web2.0 services. It reflects that the Web2.0 service industry is influenced by a few companies.

Among the 47 subgroups that are created through the type of service criterion, 34 subgroups provide less than 5 Web2.0 services. Only one subgroup consists of more than 20 Web2.0 services (Fig. 5).

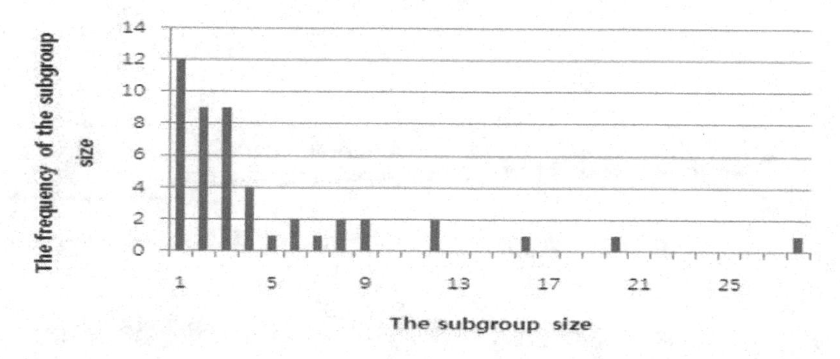

Fig. 5. Distribution of the subgroup size for subgroups based on the type of service criterion

Fig. 5 illustrates that the distribution of Web2.0 service is also uneven for the type of service criterion. A majority of Web2.0 services belong to a few service types (e.g., Map Services and Shopping Services). It mirrors that Web2.0 system is constituted through a few focal services (e.g., Mapping Service) which are supported through peripheral services.

4.2 Trend Analysis of Relative Relationship Values

In order to illustrate the trend of the relative relationship values across time, we calculate the relative relationship values for the self, internal, and external relationships for each month during the period from September 2005 to May 2007. The relative relationship values that are shown in Fig. 6 and Fig. 7 for each month are based on the cumulated data up to the month shown. For example, the self relationship value in July 2006 is the total number of self relationship in the Web2.0 service network from September 2005 to July 2006.

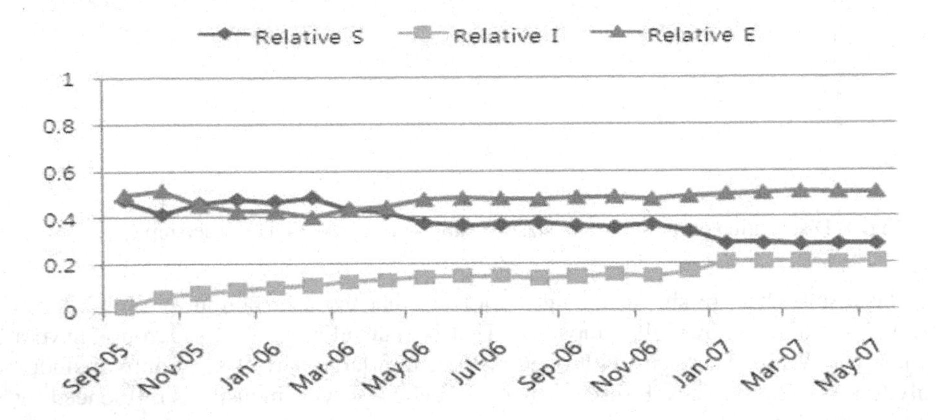

Fig. 6. The trend of relative relationship values calculated for the self, internal, and external links with respect to the ownership criterion

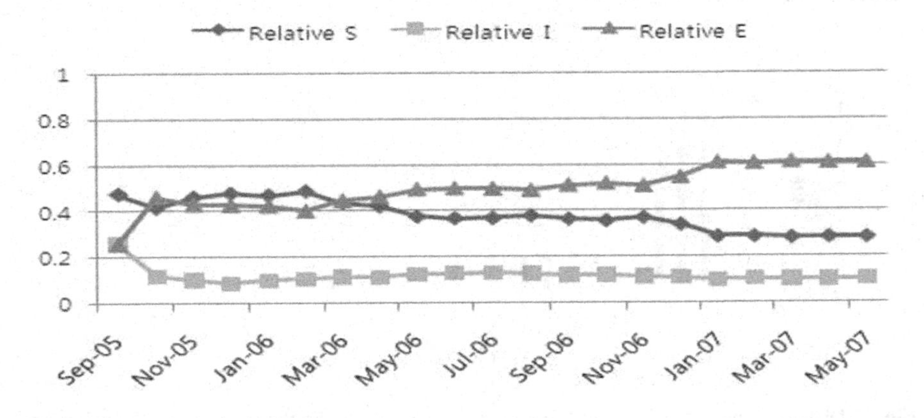

Fig. 7. The trend of relative relationship values for the self, internal, and external links with respect to the type of service criterion

With respect to the ownership criterion, Fig. 6 shows that the relative relationship value of the internal links increases steadily (The curve is named *Relative I* in Fig. 6). This means that the number of mashups developed with Web2.0 services belonging to a same subgroup increases over time. However, it is low during the entire period, and reaches only a value of about 0.2 at the end of the observation period in January 2007. The relative relationship values of the self and external links are equal during the initial period (The curves are named *Relative S* and *Relative E* in Fig. 6). In the later period, however, the relative relationship value of the external link is superior. This implies that the Web2.0 service network is open in the later periods. In particular, we can state that there are Web2.0 services of different ownership subgroups, which are not bound by the ownership of the Web2.0 service.

With respect to the type of service criterion, the comparison of the relative relationship values shows that the relative relationship value of the external links is superior and dominant at the end of the observation period (The curve is named *Relative E* in Fig. 7). Furthermore, the relative relationship value of the self links decreases after being superior in the initial period (The curve is named *Relative S* in Fig. 7). This trend shows that the structure of the Web2.0 service network is independent of the types of services. The Web2.0 service network evolves such that it destructs the barriers of service types.

4.3 Analysis of Enhanced-EIS-Indexes

The Enhanced-EIS_r-Indexes for the Web2.0 service network under both subgroup criteria, the ownership criterion and the type of service criterion, are about - 0.12 for the period of May 2007 (Fig.8).

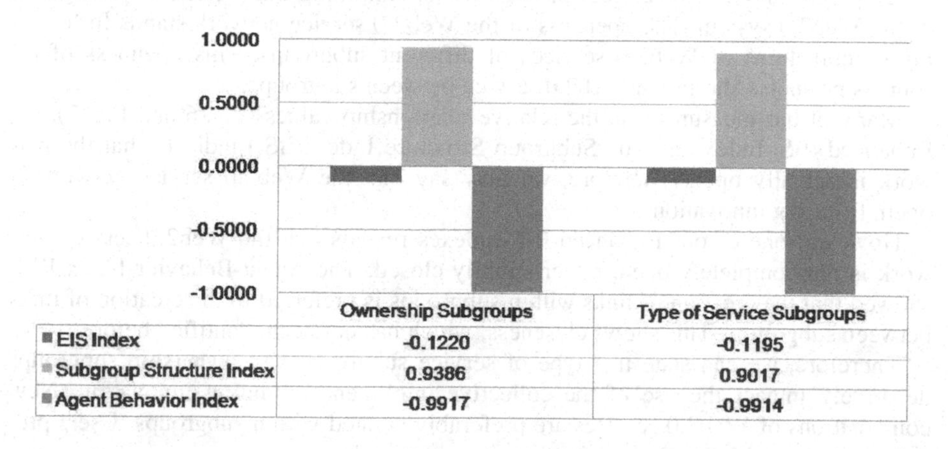

	Ownership Subgroups	Type of Service Subgroups
EIS Index	-0.1220	-0.1195
Subgroup Structure Index	0.9386	0.9017
Agent Behavior Index	-0.9917	-0.9914

Fig. 8. The Enhanced-EIS_r-Indexes, the Subgroup Structure Indexes EIS_s, and the Agent Behavior Indexes EIS_a for the Web2.0 service network of May 2007 with respect to the ownership criterion and the type of service criterion

The Subgroup Structure Indexes EIS_s for both subgroup criteria are very high, namely 0.9386 and 0.9017. It shows that the Web2.0 service industry is organized such that new Web2.0 mashups can be created through the combination of existing Web2.0 services that belong to different subgroups. It exhibits openness.

However, looking at the Agent-Behavior-Index EIS_a for the ownership criterion and the type of service criterion, we can see that they are -0.9917 and -0.9914. That implies that Web2.0 services have been used to establish relatively more links within subgroups than across subgroups. It suggests that users are captivated within subgroups. This complies with the theory that agents are more likely to make relationships with colleagues of the same subgroup, with who they share a common culture and a strong communication channel [16]. Similarly, we can suspect that a user, who is familiar with a service type or with the programming interface of a particular company, may tend to apply Web2.0 services of the same service type or the same company. In both cases, the cost for developing new Web2.0 services is low, since the user can reuse his/her existing knowledge about a type of service and programming interfaces of a company, respectively.

Consequently, these results show that the Web2.0 service network is not as open as expected and suggested by the comparison of the relative relationship values of Section 4.2. The Enhanced-EIS-Indexes, that have been introduced, helped identifying that there are preferences for establishing links within the same subgroup. This also explains the low EIS_r value of the Web2.0 service network ($EIS_r = -0.12$), revealing that the Web2.0 service network is even slightly closed.

5 Discussion of the Openness of the Web2.0 System

In this paper, the Web2.0 service network was defined as a network of innovation activities impacted by subgroups of Web2.0 services. The network is driven by the development of Web2.0 services through users, exhibiting the collective intelligence of the Web2.0 system. The openness of the Web2.0 service network stands for creating combinations of Web2.0 services of different subgroups. This openness of resources promotes sharing of Web2.0 service between subgroups.

Many of our measures (i.e. the relative relationship values (Fig. 6 and Fig. 7), the Enhanced-EIS_r-Index, and the Subgroup Structure Index EIS_s) indicate that the network is actually open. Therefore, we may say that the Web2.0 service network is open, fostering innovation.

However, one of our Enhanced-EIS-Indexes reveals that the Web2.0 service network is not completely open, rather slightly closed. The Agent-Behavior-Index EIS_a showed that the creation of links within subgroups is preferred to the creation of links between subgroups. This shows closeness, which has not been identified before.

Therefore, we can state that type of service subgroups and ownership subgroups negatively impact the use of the collective intelligence, limiting innovation. New compositions of Web2.0 services are preferably created within subgroups. Users prefer subgroups, with which they are already acquainted.

In the future, we will extend our research by considering the original Krackhardt and Stern Index in conjunction with the Enhanced-EIS-Indexes. In addition to this, an evolutionary model, which considers the behavior of agents and the impact of subgroups, is planned to be simulated.

References

1. O'Reilly, T.: What is Web2.0: Design Patterns and Business Models for the Next Generation of Software. Communications & Strategies 65, 17–37 (2007)
2. Ferris, C., Farrel, J.: What are Web Services? Communications of the ACM 46 (2003)

3. Gloor, P., Cooper, S.: Coolhunting: Chasing Down the Next Big Thing. AMACOM, New York (2007)
4. Hwang, J., Altmann, J., Kim, K.: The Structural Evolution of the Web2.0 Service Network. Online Information Review 33, 1040–1067 (2009)
5. Castilla, E.J., Hwang, H., Granovetter, E., Granovetter, M.: Social Networks in Silicon Valley. In: Lee, C.M., Miller, W.F., Hancock, M.G., Rowan, H.S. (eds.) The Silicon Valley Edge: A Habitat for Innovation and Entrepreneurship. Stanford University Press, California (2000)
6. Chesbrough, H.W.: Open Innovation: The New Imperative for Creating and Profiting from Technology. Harvard Business School Press, Boston (2003)
7. Gawer, A., Cusumano, M.A.: Platform Leadership: How Intel, Microsoft, and Cisco Drive Industry Innovation. Harvard Business School Press, Boston (2002)
8. Henkel, J.: Selective Revealing in Open Innovation Processes: The Case of Embedded Linux. Res. Policy 35, 953–969 (2006)
9. Ethraj, S.K.: Allocation of Inventive Effort in Complex Product Systems. Strategic Management Journal 28, 563–584 (2007)
10. Kim, K., Altmann, J., Hwang, J.: The Impact of the Subgroup Structure on the Evolution of Networks - An Economic Model of Network Evolution. NetSciCom. In: IEEE Intl. Workshop on Network Science for Communication Networks, IEEE Infocom 2010, USA (2010)
11. Scott, J.: Social Network Analysis: A Handbook. Sage Publication, London (1991)
12. Wasserman, S., Faust, K.: Social Network Analysis: Methods and Applications. Cambridge University Press, Cambridge (1994)
13. Burt, R.S.: Brokerage and Closure: An Introduction to Social Capital. Oxford University Press, Oxford (1994)
14. Lin, N.: Social Capital: A Theory of Social Structure and Action. Cambridge University Press, Cambridge (2001)
15. Walter, J., Lechner, C., Kellermanns, F.W.: Knowledge Transfer between and within Alliance Partners: Private Versus Collective Benefits of Social Capital. Journal of Business Research 60, 698–710 (2007)
16. Krackhardt, D., Stern, R.N.: Informal Networks and Organizational Crises: An Experimental Simulation. Social Psychology Quarterly 51, 123–140 (1988)
17. Müller-Prothmann, T., Siegbert, A., Finke, I.: Inter-Organizational Knowledge Community Building: Sustaining or Overcoming Organizational Boundaries? Journal of Universal Knowledge Management 0, 39–49 (2005)
18. Julsrud, T.E.: Core/Periphery Structures and Trust in Distributed Work Groups: A Comparative Case Study. Structure and Dynamics: eJournal of Anthropological and Related Science 2, 1–30 (2007)
19. Flew, T.: New Media: An Introduction. Oxford University Press, Melbourne (2008)
20. Tapscott, D., Williams, A.D.: Wikinomics: How Mass Collaboration Changes Everything. Penguin Group, USA (2008)
21. Feiler, J.: How to do Everything with Web2.0 Mashups. McGraw-Hill, New York (2008)
22. Yee, R.: Pro Web2.0 Mashups: Remixing Data and Web Services. Apress (2008)
23. Lévy, P.: From Social Computing to Reflexive Collective Intelligence: The IEML Research Program. Information Sciences 180, 71–94 (2010)

Collective Intelligence in Teams – Practical Approaches to Develop Transactive Memory

Michael W. Busch and Dietrich von der Oelsnitz

Technische Universität Braunschweig, Institut für Organisation und Führung, Abt-Jerusalem-Str. 4, 38106 Braunschweig

Abstract. The socio-cognitive approach to teamwork has gained a lot of attention recently. Especially the concept of transactive memory (i.e., knowledge about each other's knowledge) has been fruitfully applied to the team level. First, we extend the concept of transactive memory by considering a wider range of interpersonal aspects (e.g., personal traits, external relations, background knowledge). Second, we delineate practical approaches to develop transactive memory quickly. We distinguish between two training sequences: knowledge disclosure and knowledge updating. Whereas cross-training is an appropriate training approach at the beginning of teamwork, we refer to the after action review as an effective tool to update knowledge about each other in ongoing teamwork activities. Finally, open questions are discussed.

Keywords: transactive memory, cross-training, after action review.

1 Introduction

Teams have become a cornerstone of modern organizations. Bringing together experts from different domains can help to create synergies and to put forth new ideas. Innovative teamwork is widely viewed as a means to effectiveley combine individual knowledge, skills, and abilities (KSA) (e.g., to develop new products, to restructure organizations, or to foster the quality of strategic decision making).

We use and extend the concept of transactive memory, which describes the knowledge about who knows what within a team setting. Transactive memory has been identified as an essential prerequisite for teams to cooperate efficiently. We first present a holistic conception of transactive memory and then show how to develop this kind of knowledge quickly. The latter question has not gained the attention it deserves so far. Thus, we want to deepen our understanding not only about the concept of transactive memory but also about the practical mechanisms underlying the development of transactive memory, for it is not sufficient to simply strengthen communication efforts to achieve high levels of transactive knowledge. On the contrary, there is a need for an elaborate training program in high-performance teams.

Thererfore we distinguish between two phases or training sequences, which are both necessary to develop und to maintain transactive memory in teams: knowledge disclosure and knowledge updating. Finally, we identify key directions for future research.

T.J. Bastiaens, U. Baumöl, and B.J. Krämer (Eds.): On Collective Intelligence, AISC 76, pp. 107–119.
springerlink.com

2 Extending the Concept of Transactive Memory in Teams

Originally, psychologist Daniel M. Wegner used the concept of transactive memory to explain patterns of behavior in close relationships [1]. Whereas the husband is usually responsible for technical issues such as repairing the car or mowing the lawn, the wife is responsible for networking, planning anniversaries and so on. As a result there is an often unquestioned division of labor among partners. They rely upon each other's special competencies. Hence, it is not necessary to know or to learn everything. To put it on another way: each partner has detailed knowledge in his or her domains of expertise, but the knowledge about the domains of expertise of his or her partner can be thought of as meta-knowledge. This knowledge about the knowledge of the partner, however, is essential to coordinate behavior and to jointly solve problems.

Applied to the team level, transactive memory has become a key concept to describe and to explain coordination processes and information behavior in general. Transactive memory helps team members to assess whom to ask, to whom to pass information, and how to evaluate incoming information. This is especially useful under high work load conditions because team members hold shared expectations which in turn reduce the amount of communication necessary to adjust each other. Team members are "aware of situations in which individual teammates may require assistance and to anticipate what type of assistance those teammates prefer" [2].

According to Wegner transactive memory is primarily knowledge about team members' domains of expertise [1]. Certainly, expertise is crucial to understand collective behavior in teams concentrated on common goals. But there are other aspects of interpersonal knowledge which should be considered as well. Member familiarity, e.g., is an important concept related to the notion of transactive memory. It describes knowledge about eacht other's "preferences, habits, and values" [3]. Not only does friendship or prior contact lead to more effective coordination processes between teammates, but it may also be related to "an increase in requests for and acceptance of backup" [4]. Following these results, we think that transactive memory should be defined more broadly to include other facets of personal traits and patterns of behavior. So we propose five core elements of transactive memory:

- Knowledge about knowledge, skills, and abilities: This knowledge fosters an efficient information processing in teams: "The study of transactive memory is concerned with the prediction of group (and individual) behavior through an understanding of the manner in which groups process and structure information" [1]. To know who possesses a respective talent and how to use that talent strengthens not only the quality, but also the speed of team decision making. Transactive memory is embedded in a collective system of encoding, storing, and retrieving information. Both – organizations and teams face similar challenges, i.e. they have to establish common rules for managing their knowledge effectively, and they have to define standards for orchestrating their communication and interaction processes. Knowledge about KSA is pivotal in this context. It serves as an implicit coordination mechanism which has been defined as "the ability of a team to act in concert by predicting the needs of the task and the team members and adjusting behavior accordingly, without the need for overt communication" [5].

- Knowledge about nominal characteristics such as gender, status, or race: This knowledge arises automatically when people meet first. Because this knowledge is created more or less accidentally, it is normally too superficial to lead us to accurate conclusions on people: "The simplest, and often most inaccurate, way to form directories is stereotyping" [6]. To avoid that this knowlede becomes the only source about one another, geographically dispersed team members, for instance, should meet face-to-face from time to time, especially early in their life cycle.
- Knowledge about personal traits: Personal traits refer to the style or manner in which people interact with their environment, e.g., how they deal with criticism, how they act under stress, how they learn from defeat, or how they are willing to admit to mistakes. This knowledge is very important to work together smoothly and to create an atmosphere of psychological safety [7]. It helps to avoid relational conflicts between team members which have been shown to be detrimental to workgroup performance [8]. Personal traits as part of transactive memory have been largely ignored in comparison to KSA. Nevertheless, they are intensively discussed in conjunction with questions about team composition. Consequently, it makes sense to integrate these findings within the framework of transactive memory research, because it is not only important to know what others know but also how they use their knowledge daily. The "big five" factors of personality can provide a basis for further research findings. These factors are openness, conscientiousness, extraversion, agreeableness, and neuroticism [9].
- Knowledge about team members' personal background: This knowledge stems from the fact that the western culture does not value silence. Silence makes people feel uncomfortable. In periods of low work load, during lunch time, or when people greet one another they talk about their families, their hobbies, their activities during the last weekend, or other personal aspects of their life (e.g., their mood and feelings or their well-being). Although this knowledge is not really necessary for attaining the team's goal, it – nevertheless – fulfils certain functions for the team as it contributes to a positive and supportive working culture. It should be mentioned, however, that sharing this kind of knowledge is voluntary in nature, i.e., team members cannot be coerced to share.
- Knowledge about team members' social capital: This knowledge is useful to gain knowledge from outside of the team. It can be defined as "shared awareness of group member external ties" [10]. As team members normally belong to many other groups (e.g., communities of practice or any other kind of social networks) this knowledge can be used to widen the team's horizon: "An effective transactive memory system should have bridges spanning structural holes such that information can flow from each group within the network to every other group" [11]. If problems arise which the team itself is unable to resolve these external ties become especially valuable as they extend the team's knowledge base and probably also its innovative power. Some authors assume that the external orientation accounts for the success of outstanding teams. Teams embedded in dense networks of contacts inside and outside the

organization "develop a clear understanding of the environment (scouting), build support with key executives (ambassadorship), and coordinate with other groups that can contribute to their project (task coordination)" [12].

If we integrate all these elements we will get to a more comprehensive view of transactive memory (cf. Fig. 1). It is not only important to consider hard facts of interpersonal knowledge, such as domains of expertise, but also to look at more soft facts, such as personal traits and background, which may invoke feelings of friendship and solidarity among team members. Topical research findings from Ferriani, Corrado and Boschetti can help to clarify this view. In their longitudinal study on the U.S. feature film industry they found that many directors collaborate with a stable nucleus of key members over subsequent projects: "In a context constantly in flux, due to the free-lance market-based organization of labor and the short-term nature of the production process, reiterated collaborations provide stability and continuity" [13]. Whether driven by friendship or by functional demands, these strong ties are certainly characterized by high levels of transactive or idiosyncratic knowledge about each other. This knowledge reduces transaction costs, i.e. it shortens the time needed to get along with one another or, as the late Sydney Pollack put it: "When you find people you can work with you never want to give them up" [13].

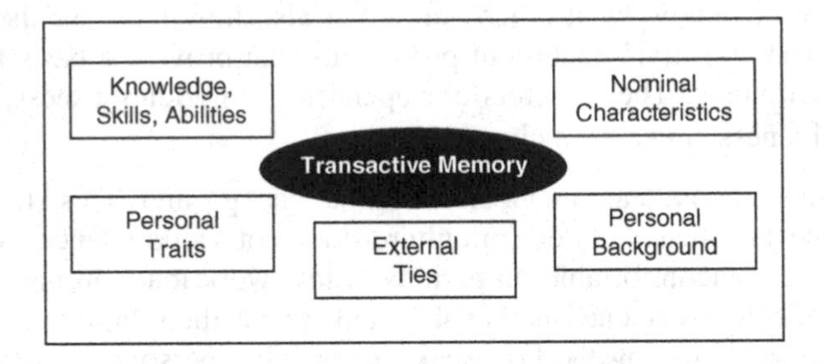

Fig. 1. Holistic Conception of Transactive Memory [14]

3 The Development of Transactive Memory

It has to be mentioned first that there is no panacea to develop transactive memory. On the one hand, there are many factors which influence the need for transactive memory (e.g., different levels of task interdependence, diversity and longevity of a team, team member stability, trust, and the willingness to share information), on the other hand transactive memory is a complex theoretical construct. Hence, training measures cannot be reduced to a single method but should encompass different aspects of interpersonal and interpositional knowledge. Another hurdle to overcome can be found in structural constraints (e.g., budgetary restrictions, time pressure, and an urge for results) which often prevent sophisticated training programs or workshops for newly formed teams.

Moreover, scientists are unanimous in their recommendation to install measures for team building right at the beginning to be effective. Some authors, however, question that view argueing "that team skills develop naturally through time spent together, implying that specific team process training is not required" [15]. We state, however, that such unplanned development makes initial coordination processes more error-prone, triggering intragroup conflicts and slowing down work progress. Above that, the less systematic training measures are, the more it takes time to get to a teamwide mutual understanding. In Tuckman's words: Training teams at the beginning helps them to get soon into the performing stage, i.e. to avoid the storming stage [16].

While there is consensus on when to apply training methods and on how these training methods should be designed (i.e., normally team-based, not individual-based), there is no consensus on the question which training methods are really appropriate to develop transactive memory yet. Certainly, asking teams to communicate as often as possible cannot be questioned. The same is true for collaborative information technologies which have become an indispensable part of teamwork: "The learning channels or communication media used to convey transactive knowledge play a crucial role in facilitating knowledge transfer" [17]. In an overarching analysis Kozlowski and Ilgen, however, conclude that "beyond familiarity, shared experience, and face-to-face-interaction, the research base to help identify techniques for enhancing transactive memory is as yet not sufficiently developed to warrant specific recommendations for how to enhance it in teams. This is an obvious target for vigorous and rigorous research" [18]. They propose that "the use of interpositional cross-training, which has proven useful in the development of shared mental models, may also help to foster the development of better transactive-memory systems" [18]. Following this proposal, we will take a closer look at cross-training as a means to develop transactive memory at the beginning of teamwork.

3.1 Knowledge Disclosure through Cross-Training

As mentioned before, cross-training is not a single method but more an umbrella term capturing different training approaches and training levels. It consists of positional clarification, positional modeling, and positional rotation. Figure 2 shows how these training modules are connected.

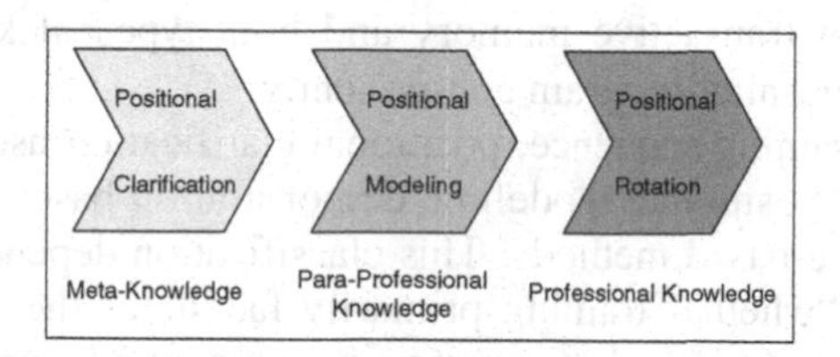

Fig. 2. The Three Modules of Cross-Training [14]

Orignially designed for developing interpositional knowledge, cross-training has the potential to develop interpersonal knowledge as well. Especially in project-based work, roles and responsibilities are not that fixed as they are in an organizational, more hierarchical setting. They have to be much more flexible to adapt to environmental changes. It makes sense to combine interpositional knowledge with elements of interpersonal knowledge. To know what others have to do is nearly congruent with the knowledge about what they have to *know* on their assigned job.

The three types of training should be seen as a sequence. The complexity and depth of knowledge increases from module to module. Whereas positional clarification and positional modeling give only a superficial impression of team members' job requirements (which should not be confused with actual mastery), the question whether positional rotation is possible at all depends on the cognitive distance between team members' specialties: "It would be overly optimistic to think that a designer could easily educate an engineer to understand design with the same level of sophistication that the designer has gained after years of training and experience" [19]. So it is neither feasible nor necessary to fully cross-train all team members. The choice depends, first and foremost, on the team's task: "The prototypical work team from a manufacturing setting often has people cross-trained on a full set of team skills. Full cross-training is inappropriate for most knowledge-work teams, however, because the in-depth, specialized knowledge required is expensive to obtain (...) Teams require members to have, at a minimum, enough understanding of the skills of their teammates to be able to discuss issues and trade-offs as the team goes through the cycle of considering divergent views and arriving at convergence on a direction. Familiarity across disciplines provides a basis for communicating across the thought-worlds of the different disciplines" [20]. Cross-training in knowledge-work teams aims to develop – what Marco Iansiti called – T-shaped skills which he conceives to be essential for any type of integration team involved in the R&D process. Members possessing a T-shaped combination of skills "are not only experts in specific technical areas but also intimately acquainted with the potential systemic impact of their particular tasks. On the one hand, they have a deep knowledge of a discipline like ceramic materials engineering, represented by the vertical stroke of the T. On the other hand, these ceremic specialists also know how their discipline interacts with others, such as polymer processing – the T's horizontal top stroke" [21]. As claimed before, these more interpositional aspects of knowledge should be amalgamated with the more interpersonal aspects of knowledge. Both types of knowledge can be found in our holistic conception of transactive memory and both types of knowledge should be integrated in a cross-training program consequently.

Looking at each training sequence, positional clarification uses primarily information-based methods, positional modeling demonstration-based methods, and positional rotation practice-based methods. This classification depends on how one delivers the training, i.e. "whether training primarily facilitates the delivery of concepts, facts, knowledge, or theories – *information based methods*; or illustrates by visual behaviors, actions or strategies to be learned – *demonstration-based methods*; or whether training allows the trainee hands-on practice and provides feedback on progress – *practice-based methods*" [22].

As the words express, *positional clarification* wants to create transparency about team member's knowledge. In self-managing work teams (e.g., in manufacturing or in service settings) positional clarification unfolds the variety of skills of each team member. Figure 3 shows a simplified skill chart of a work team.

Machine	Operator		
	1	2	3
Lathe	•		○
Lathe	•		○
Hobber	○	•	
Hobber	○	•	
Cylindrical Grinder		○	•
Cylindrical Grinder		○	•
• operator's main responsibility ○ operator shares whenever possible			

Fig. 3. Simplified Skill Chart of a Work Team [23]

In knowledge-work teams positional clarification helps to raise awareness about each other's position. A workshop preceding teamwork may help to clarify interpositional and interpersonal knowledge. In such workshops sheets containing individual competency profiles can be handed out and discussed or team members can talk about their professional background and their experiences. It leads to the general question on how knowledge can be made visible. Knowledge about one another may either be transferred from team member to team member or by using codified knowledge like knowledge maps. Knowledge maps are graphic directories of knowledge "-sources (i.e., experts), -assets (i.e., core competencies), -structures (i.e., skill domains), -applications (i.e., specific contexts in which knowledge has to be applied, such as a process), or -development stages (phases of knowledge development or learning paths)" [24].

A facilitator may help the team to understand the general importance of team transactive memory und put it in the context of shared mental models of which transactive memory is an essential part of. Shared mental models in teams are common or overlapping cognitive representations of the team's reality: "[O]ne that describes the equipment (equipment model), one that describes the task (task model), and two that describe the team – one that describes the roles, responsibilities, and interactions of team members (team interaction model) and one that describes the team members themselves (team model)" [25]. The facilitator should be supported by a skilled and experienced project manager who can give a more vivid picture of what the team is going to expect in daily work.

Compared to positional clarification, the sequence of *positional modeling* is less theoretical, i.e. positions or team members' roles are observed in their natural setting. The tasks of another teammate may be either observed in a simulated situation or in real work situations. We think that it could be beneficial to use think-aloud techniques to accelerate mutual understanding and to foster the explication of tacit knowledge. Ericsson and Simon [26] distinguish between concurrent and retrospective reporting.

The retrospective report is given by the subject immediately after the task is completed. Compared to concurrent reporting the process that is completed, however, cannot be altered and influenced any more. In addition to think-aloud techniques more indirect ways of observing teammates' behavior can be used (e.g., shadowing), similar to a mentor/mentee relationship. All these methods are labor and time-intensive.

Finally, *positional rotation* is experientially based training i.e., team members actually perform all of the duties or different parts of the duties of their teammates. "This method is similar to job rotation in that team members gain first-hand knowledge and experience in the specific tasks of others (...) Ideally, team members would be trained in those tasks that demand cooperation and high interdependency among teammates" [27]. Within this training sequence an overall task should be decomposed into components to be separately trained: "*Segmentation* involves breaking up a task into a temporal series of stages and designing part-task training around these individual stages. *Simplification* involves making the task initially easier to perform in some manner (...) prior to having the trainee move to the full version of the task (...) *Fractionation* involves decomposing a complex task into a set of individual activities that are performed in parallel in the overall task" [28]. Apart from these general questions concerning the appropriate design of positional rotation, Salas and Cannon-Bowers point out that "to be effective, practice needs to be guided by cuing, feedback, coaching, or any other mechanism that helps the trainee to understand, organize, and assimilate the learning objectives" [22].

3.2 Facilitating Knowledge Updating through After Action Reviews

The development of situation awareness (SA) becomes crucial when a team starts working. Individual outputs need to be integrated, interdependencies to be identified, and the team as a whole has to adapt to changing conditions. "Most simply put, SA is knowing what is going on around you. Inherent in this definition is a notion of what is important" [29]. Team situation awareness can be understood as a dynamic shared conception of the changes taking place within and outside the team. These changes include changes of team members' duties and competencies as well as changes of the team's constituencies or other stakeholders. A constant flow of information is needed to maintain situation awareness. This is not only a technological question but also a question concerning the team's behavior and interaction processes. As individual progress is directly linked to the team's progress, a team should install mechanisms to ensure that knowledge, ideas, and observations of team members are shared within the team. Thus, teams should learn how to provide feedback effectively.

One device to facilitate such feedback processes is the use of the after action review which has proven an effective tool in the U.S. Army. At a first glance the after action review seems quite trivial, because it uses only four questions again and again to review events and to reflect upon measures that should be taken to improve the team's performance (cf. Figure 4) [30].

But if we take a closer look at feedback processes, we will find that each team has to learn it from scratch, because "[d]etecting and correcting error is psychologically non-trivial as it involves loss of face and possible loss of confidence" [32]. So the team leader has to play the role of a facilitator who initiates and monitors feedback

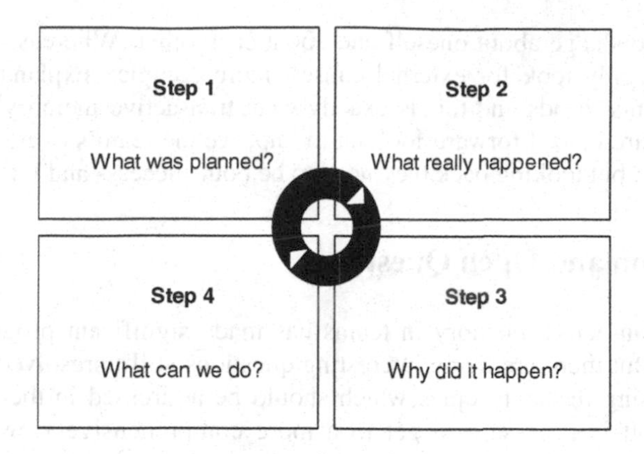

Fig. 4. The After Action Review as a Learning Cycle [31]

processes in ongoing after action reviews. Collison and Parcell provide practical advice [33]:

- Focus on the process rather than the content.
- Watch people's body language – it will tell you more than their words alone.
- Ensure a balanced contribution from all staff – ask questions of the quiet ones.
- Trust your instinct to ask the 'unasked questions'.
- Clarify distinctions between facts and opinions.
- Get the participants to focus on what actions they will take, rather than on what others will do.

Accordingly, the after action review is not something that can be imposed on team members, instead it "is a characteristic that team members nurture over time through a variety of self-development processes" [34]. A team has to learn how to use the after action review effectively. Transactive memory can contribute to accelerate this learning process, because the more we know about each other, the better we can learn from each other, and the more we are willing, to give and to accept feedback from one another. So transactive memory is both a precondition of successful after action reviews, and a consequence of after action reviews, because these reviews deepen and broaden the knowledge about each other.

In order to gain knowledge about changes in each other's domains of expertise quickly, critical incidents are highly important because they obviously unfold individual strengths and weaknesses [35]. They result in feelings of frustration or happiness. Thus, attention of participants is focussed when analyzing these incidents during an after action review session. It is only human to concentrate on mistakes and failures, yet successes should be analyzed, too, in order to replicate them. Ellis and Davidi showed for the Israel Defense Forces that explanations of successes at the beginning of a training session were rather simple and straightforward, but over time "participants' mental models of successful events became increasingly more complex and included increasingly more causal explanations that did not reflect situational reasons but instead reflected issues relating to their prior knowledge and task planning" [36]. To ask team members to reflect upon especially successful events therefore may lead to more fine-grained knowledge about oneself and about each other. Whereas simple explanations of events only look for external causes, more complex explanations also look inside participants' heads and this is exactly what transactive memory is about. After action reviews are indeed forward-looking to improve the team's overall competencies and interactions, but looking back they need to be both success- and failure-driven.

4 Conclusion and Open Questions

Research on transactive memory in teams has made significant progress during the last ten years. But there are many interesting questions still unresolved. A short overview of promising research topics which should be addressed in the near future are among others: It is necessary to get to a more comprehensive view of transactive memory as we have pointed out before; this comprehensive view itself needs to be integrated in an even more comprehensive view, i.e., results from transactive memory research must be integrated with results from the research on shared mental models in teams. "[S]ocial sharedness is central to understanding group decision-making, provides a tie between past and current group research, and can serve a unifying function for future endeavors" [37]. Thus, future research should examine the linkage between different components of shared mental models (of which transactive memory is certainly one of the most important ones), as well as the amount of commonality or the degree of integration necessary for each component [38]: "Teams do need to share some overlapping knowledge in order to coordinate their actions and perform well. What we do not know is how much knowledge must be overlapping, and how much specialization is too much" [39].

To choose the right level of analysis is another question to be solved. Transactive memory is certainly especially insightful at the team level, but it could also be applied to the organizational level [40]. Combined with social network theory [41] it provides the basis for analyzing the roots of the knowledge creating company. It allows the organization to know what it knows [42]. Transactive memory is, metaphorically speaking, a kind of human web search engine. This personalized meta-knowledge supplements technical web search engines like Google, Yahoo, Bing, or any other kind of intranet solutions. Whereas the latter give quick answers to knowledge requests, a holistic conception of transactive memory contains elements such as intuition, mental associations, moods, or knowledge about human nature, which help to evaluate individual achievement levels, making staffing decisions more accurately, and advancing the organization's information processing capacities in general. This aspect is of high importance to the inter-organizational level, too. Strategic networks face even more information-based as well as interpersonal challenges. In summary, multi-level approaches to transactive memory seem most promising.

But transactive memory research is not only confronted with theoretical questions. Practical issues should receive much more attention in future research. Advances in training research are to be considered (e.g., blended learning, Web 2.0 technologies such as Wiki systems or tagging) to make the design of transactive memory training more effective and more interactive in nature. The new generation of technologically socialized people will call for collaborative, non-hierarchic ways to develop transactive memory. First steps in this direction have already been taken [43]. The role of technology also comes to the fore in virtual teams whose members work across boundaries (e.g., organizational, industrial, or cultural) and correspondingly have to cope with diversity in a more effective way. Finally, membership dynamics must be taken into consideration, because many teams are characterized by a high rate of turnover as functional demands change. In these cases "team training should focus on developing knowledge and skills that are transportable and that do not need to be repeatedly retrained and unlearned" [44]. If team composition often changes, management should at least make sure that there will be a stable nucleus of team members who know each other very well. Gratton and Erickson, e.g., discovered "that when 20% to 40% of the team members were already well connected to one another, the team had strong collaboration right from the start" [45]. This "transactive" nucleus accelerates team development and serves as a behavioral model for newcomers.

Researchers on transactive memory should feel much more obliged to cross their disciplinary boundary. The concept of transactive memory is predisposed to be analyzed from different angles. Neuroscience, e.g., would advance our understanding in this field tremendously. Therefore, it is not only important to know what is known in your own research field, but also to know what other research fields have to offer in order to deepen our knowledge about the complex mechanisms underlying the evolution of collective intelligence in teams.

References

1. Wegner, D.M.: Transactive Memory: A Contemporary Analysis of the Group Mind. In: Mullen, B., Goethals, G.R. (eds.) Theories of Group Behavior, pp. 185–208. Springer, New York (1987)
2. Smith-Jentsch, K.A., Baker, D.P., Salas, E.: Cannon-Bowers, J.A.: Uncovering Differences in Team Competency Requirements: The Case of Air Traffic Control Teams. In: Salas, E., Bowers, C.A., Edens, E. (eds.) Improving Teamwork in Organizations, pp. 31–54. Lawrence Erlbaum Associates, Mahwah (2001)
3. Rockett, T.L., Okhuysen, G.A.: Familiarity in Groups: Exploring the Relationship between Inter-Member Familiarity and Group Behavior. Research on Managing Groups and Teams 4, 173–201 (2002); cf. also Gruenfeld, D.A., Mannix, E.A., Williams, K.Y., Neale, M.: Group Composition and Decision Making: How Member Familiarity and Information Distribution Affect Process and Performance. Organizational Behavior and Human Decision Processes 67, 1–15 (1996)
4. Smith-Jentsch, K.A., Kraiger, K., Cannon-Bowers, J.A., Salas, E.: Do Familiar Teammates Request and Accept More Backup? Transactive Memory in Air Traffic Control. Human Factors 51, 181–192 (2009)
5. Rico, R., Sánchez-Manzanares, M., Gil, F., Gibson, C.: Team Implicit Coordination Processes: A Team Knowledge-Based Approach. Academy of Management Review 33, 163–184 (2008)
6. Peltokorpi, V.: Transactive Memory Directories in Small Work Units. Personnel Review 33, 446–467 (2004)
7. Edmondson, A.: Psychological Safety and Learning Behavior in Work Teams. Administrative Science Quarterly 44, 350–383 (1999)

8. Jehn, K.A., Mannix, E.A.: The Dynamic Nature of Conflict: A Longitudinal Study of Intragroup Conflict and Group Performance. Academy of Management Journal 44, 238–251 (2001)
9. Barrick, M.R., Stewart, G.L., Neubert, M.J., Mount, M.K.: Relating Member Ability and Personality to Work-Team Processes and Team Effectiveness. Journal of Applied Psychology 83, 377–391 (1998)
10. Austin, J.R.: Knowing What and Whom Other People Know: Linking Transactive Memory with External Connections in Organizational Groups. In: Academy of Management Best Paper Proceedings, pp. F1–F6 (2000)
11. Garner, J.T.: It's not What You Know: A Transactive Memory Analysis of Knowledge Networks at NASA. Journal of Technical Writing and Communication 36, 329–351 (2006)
12. Ancona, D., Bresman, H., Caldwell, D.: The X-Factor: Six Steps to Leading High-Performing X-Teams. Organizational Dynamics 38, 217–224 (2009)
13. Ferriani, S., Corrado, R., Boschetti, C.: Organizational Learning under Organizational Impermanence: Collaborative Ties in Film Project Firms. Journal of Management and Governance 9, 257–285 (2005)
14. Busch, M.W.: Kompetenzsteuerung in Arbeits- und Innovationsteams. Gabler, Wiesbaden (2008)
15. Prichard, J.S., Ashleigh, M.J.: The Effects of Team-Skills Training on Transactive Memory and Performance. Small Group Research 38, 696–726 (2007)
16. Tuckman, B.W.: Developmental Sequence in Small Groups. Psychological Bulletin 63, 384–399 (1965)
17. Hodgkinson, G.P., Sparrow, P.R.: The Competent Organization. Open University Press, Buckingham (2002)
18. Kozlowski, S.W.J., Ilgen, D.R.: Enhancing the Effectiveness of Work Groups and Teams. Psychological Science in the Public Interest 7, 77–124 (2006)
19. Cronin, M.A., Weingart, L.R.: Representational Gaps, Information Processing, and Conflict in Functionally Diverse Teams. Academy of Management Review 32, 761–773 (2007)
20. Mohrman, S.A., Cohen, S.G., Mohrman Jr., A.M.: Designing Team-Based Organizations. Jossey-Bass Publishers, San Francisco (1995); cf. also Cohen, S.G., Bailey, D.E.: What Makes Teams Work: Group Effectiveness Research from the Shop Floor to the Executive Suite. Journal of Management 23, 239–290 (1997)
21. Iansiti, M.: Real-World R&D: Jumping the Product Generation Gap. Harvard Business Review 71, 138–147 (1993)
22. Salas, E., Cannon-Bowers, J.A.: Methods, Tools, and Strategies for Team Training. In: Quiñones, M.A., Ehrenstein, A. (eds.) Training for a Rapidly Changing Workplace, pp. 249–279. American Psychological Association, Washington (1997)
23. Hyer, N., Wemmerlöv, U.: Reorganizing the Factory. Productivity Press, Portland (2002)
24. Eppler, M.J.: Making Knowledge Visible Through Intranet Knowledge Maps: Concepts, Elements, Cases. In: Proceedings of the 34th Hawaii International Conference on System Sciences (2001)
25. Cannon-Bowers, J.A., Salas, E., Converse, S.: Shared Mental Models in Expert Team Decision Making. In: Castellan Jr., N.J. (ed.) Individual and Group Decision Making, pp. 221–246. Lawrence Erlbaum Associates, Hillsdale (1993)
26. Ericsson, K.A., Simon, H.A.: Protocol Analysis. The MIT Press, Cambridge (1993)
27. Blickensderfer, E., Cannon-Bowers, J.A., Salas, E.: Cross-Training and Team Performance. In: Cannon-Bowers, J.A., Salas, E. (eds.) Making Decisions under Stress, pp. 299–311. American Psychological Association, Washington (1998)

28. Kirlik, A., Fisk, A.D., Walker, N., Rothrock, L.: Feedback Augmentation and Part-Task Practice in Training Dynamic Decision-Making Skills. In: Cannon-Bowers, J.A., Salas, E. (eds.) Making Decisions under Stress, pp. 91–113. American Psychological Association, Washington (1998)
29. Endsley, M.R.: Theoretical Underpinnings of Situation Awareness: A Critical Review. In: Endsley, M.R., Garland, D.J. (eds.) Situation Awareness, pp. 3–32. Lawrence Erlbaum Associates, Mahwah (2000)
30. Baird, L., Henderson, J.C., Watts, S.: Learning from Action: An Analysis of the Center for Army Lessons Learned (CALL). Human Resource Management 36, 385–395 (1997)
31. von der Oelsnitz, D., Busch, M.W.: Teamlernen durch After Action Review. Personalführung 39, 54–62 (2006)
32. Ron, N., Lipshitz, R., Popper, M.: How Organizations Learn: Post-Flight Reviews in an F 16 Squadron. Organization Studies 27, 1069–1089 (2006)
33. Collison, C., Parcell, G.: Learning to Fly. Capstone, Chichester (2004)
34. Vashdi, D.R., Bamberger, P.A., Erez, M., Weiss-Meilik, A.: Briefing-Debriefing: Using a Reflexive Organizational Learning Model from the Military to Enhance the Performance of Surgical Teams. Human Resource Management 46, 115–142 (2007)
35. Flanagan, J.C.: The Critical Incident Technique. Psychological Bulletin 51, 327–358 (1954); Butterfield, L.D., Borgen, W.A., Amundson, N.E., Maglio, A.-S.T.: Fifty Years of the Critical Incident Technique: 1954-2004 and Beyond. Qualitative Research 5, 475–497 (2005)
36. Ellis, S., Davidi, I.: After-Event-Reviews: Drawing Lessons from Succesful and Failed Experience. Journal of Applied Psychology 90, 857–871 (2005)
37. Tindale, R.S., Kameda, T.: 'Social Sharedness' as a Unifying Theme for Information Processing in Groups. Group Processes&Intergroup Relations 3, 123–140 (2000)
38. McComb, S.A.: Mental Model Convergence: The Shift from Being an Individual to Being a Team Member. Research in Multi-Level Issues 6, 95–147 (2007)
39. Hinsz, V.B., Tindale, R.S., Vollrath, D.A.: The Emerging Conceptualization of Groups as Information Processors. Psychological Bulletin 121, 43–64 (1997)
40. Nevo, D., Wand, Y.: Organizational Memory Information Systems: A Transactive Memory Approach. Decision Support Systems 39, 549–562 (2005)
41. Katz, N., Lazer, D., Arrow, H., Contractor, N.: The Network Perspective on Small Groups. In: Poole, M.S., Hollingshead, A.B. (eds.) Theories of Small Groups, pp. 277–312. Sage Publications, Thousand Oaks (2005)
42. Davenport, T.H., Prusak, L.: Working Knowledge. Harvard Business School Press, Boston (1998)
43. Allan, M.B., Korolis, A.A., Griffith, T.L.: Reaching for the Moon: Expanding Transactive Memory's Reach with Wikis and Tagging. International Journal of Knowledge Management 5, 51–63 (2009)
44. Smith-Jentsch, K.A., Cannon-Bowers, J.A., Tannenbaum, S.I., Salas, E.: Guided Team Self-Correction. Impacts on Team Mental Models, Processes, and Effectiveness. Small Group Research 39, 303–327 (2008); Lewis, K., Belliveau, M., Herndon, B., Keller, J.: Group Cognition, Membership Change, and Performance: Investigating the Benefits and Detriments of Collective Knowledge. Organizational Behavior and Human Decision Processes 103, 159–178 (2007)
45. Gratton, L., Erickson, T.J.: Eight Ways to Build Collaborative Teams. Harvard Business Review 85, 100–109 (2007)

The Need Language: A Preliminary Report

A. Abramovich[1], C.Z. Xu[1], P. Guo[1], L. Wang[1], T. Qian[1], Q. Wang[2], and P. C.-Y. Sheu[2]

[1] State Key Lab of Software Engineering (SKLSE)
Wuhan University, Wuhan, China
webdao@gmail.com
[2] Department of EECS
University of California, Irvine
qiw@uci.edu

Abstract. Knowledge representation is a key task of both computing science and programming practice. Suffice it to say that any program is a knowledge representation of a certain problem solution. However till now there are no means for the representation of application problems' decision methods, for the representation of the environments making these problems, and for the representation of the communications between different knowledge environments. Today's evolution of IT requires such knowledge representation tools. This paper proposes a knowledge representation language that allows a system of knowledge to be represented in a comfortable way for wide range of users and for automatic and semi-automatic problem solving in a suitable form.

Keywords: semantics, ontology, reasoning, domain knowledge representation, Need Satisfaction Domain, constructive element, a target activity, a local target activity, resource, Need Language.

1 Introduction

In daily life people continuously solve problems. Herewith, new problems seldom occur. In most cases we deal with problems that are available in the scope of own or common experiences. By experience we mean a collection of knowledge and skills that are resulted from activities. This collection contains knowledge about domain situations and domain processes which produce these problems. Generally speaking, a decision is reduced to the detection of a problem, to the synthesis of solution using available experiences and to control of solution:

$$\text{Decision} = \text{Recognition} + \text{Synthesis} + \text{Control} \quad (1)$$

Equation (1) would be useful, if some information technologies would model (or just support) the human decision approach.

Traditional computing distinguishes two styles of programming: declarative and imperative. With the imperative style we must write rigorous instructions for the computer to follow, step-by-step. In the declarative style we just inform the computer

T.J. Bastiaens, U. Baumöl, and B.J. Krämer (Eds.): On Collective Intelligence, AISC 76, pp. 121–133.
springerlink.com

about the problems. A declarative program is essentially a specification. Using R. A. Kowalski's equation [6]

$$\text{Algorithm} = \text{Logic} + \text{Control} \tag{2}$$

It is possible to say that declarative programming describes the logic of algorithms, but not necessarily the control, whereas imperative style of programming deals with both logic and control. Herewith both these styles realize the synthesis of algorithms. That is, the process of algorithm synthesis is outside computing. Moreover, both declarative and imperative styles of programming allow describing the decisions of separated problems, which have been extracted from a situational context. Herewith the remaining domain reality is left out. In other words, it is possible to say that the representation of domain knowledge and software synthesis logic is beyond the capabilities of traditional programming.

Attempts to represent a domain environment by means of existing programming languages have been unsuccessful also for another important reason that the concept of domain knowledge is not rigorously defined. Usually, programmers rely on intuition and (domain theory) a collection of types, operations, laws, and inference rules that are structured arbitrarily. The lack of a strict definition of domain entails the absence of criteria for completeness and correctness of its description. Since any specific domain knowledge is oriented to certain need (or needs) satisfaction and consists of several various domains, we focus on the representation of the Need Satisfaction Domain that includes all necessary knowledge for satisfying the given need (or needs).

Since the decision is, mainly, a process of recognition and synthesis, Need Satisfaction Domain should be presented as data including both declarative and imperative information.

Thereby, the present paper deals with a knowledge representation language designed for the representation of domain knowledge. The domain of need satisfaction is in the form of data that includes both declarative and imperative information. The rest of the paper is organized as follows: in Section 2, we introduce work related to this paper. In Section 3 we introduce briefly the basic principles of domain knowledge representation. In section 4 we present the NL language syntax in a general form. In Section 5 we present the implementation of the NL language. Lastly, the paper ends with a short conclusion.

2 Related Work

The research on domain knowledge representation has last for many years. Some studied domain knowledge representation including rules, integrity constraints, type definitions, policy decisions and the applications of domain modeling to support specific operational goals [17]. However, it is not simply knowledge representation; applications also need knowledge reasoning and most application knowledge acquired from experts. Similar to open frame systems like RLL[22] and KODIAK[23], CreekL[18], presented by Agnar Aamod, makes an inference by property inheritance and constraint enforcement and enables a tight integration of case-specific and domain knowledge by emphasizing a through, explicit representation of all concepts. A knowledge representation system attempting to integrate case-specific and general domain knowledge usually is object-oriented and frame-based. While some people represent domain knowledge with xml on semantic web (e.g. [19]), XMLKR is an object-oriented method of xml for knowledge representation [19]. Ontology-based domain knowledge representations were introduced in some papers (e.g. [20], [21]). A kind of formalism was presented based on three-level, ontology-based knowledge representation structures (i.e., subject ontology level, knowledge concept ontology level and learning material ontology level) [20]. Another representation framework for domain knowledge based ontology was proposed in [21]. By defining class nodes and instance nodes, a description method for data nodes was presented and the class node properties are used to describe the domain knowledge formally. The system obtains results by reasoning about the semantic relations between class nodes, instance nodes and properties. However, the applicability is limited because of the lack of different domain integrations, and is not simple enough for users.

3 Domain Knowledge Representation

We consider target knowledge as the knowledge which a target system operates on to satisfy a given need. We define Environmental Knowledge as the knowledge about environmental needs.

By Environmental Knowledge we mean knowledge about the environment that both motivates and governs the existence of the target system. This knowledge about an external environment contains the cumulative knowledge of all external factors that influence the target system's life cycle or/and the knowledge about the external processes (activities) that produces the environmental needs (optional). By Target Knowledge we also mean the knowledge about the target system itself, which serves the satisfaction of Human needs and is represented by the description of the related target activities and situations. Generally speaking, by target activity we mean a human activity aimed to satisfy certain social need and operate with the target knowledge.

In the context of software synthesis the target knowledge is the knowledge about the resources of the target activities and the structures of the target activities that define the usage of their constructive elements.

Fig. 1 provides the content of a Need Satisfaction Domain in a general form. Any social need is associated with one of more ways of its satisfaction through a certain target activity. Every target activity belongs to a certain target environment and is represented by its resources, a configuration of its constructive elements and the known situations which are related to the target activity's execution process.

Any NL data destined for a Need Satisfaction Domain description belongs to the same type of data, namely, any NL data is a Reasoning Resource. Depending on domain situations the NL language allows users to qualify every reasoning resource as a missing resource (hereinafter a need), an available resource (hereinafter a resource) or a constructive resource (hereinafter a constructive element or an activity).

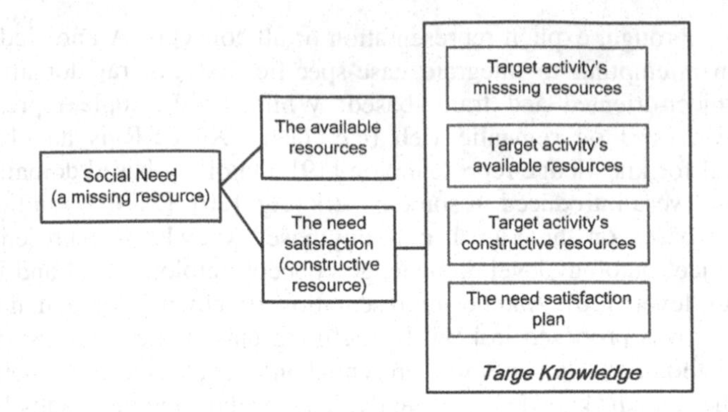

Fig. 1. Need Satisfaction Domain's content

4 Need Language Syntax

4.1 Semantics and Ontology

Any NL language sentence consists of an NL's reserved word (semantic_tag) that is separated by a colon from a list of domain concepts and/or assertions:

semantic_tag: (domain_concepts/assertions)

A domain concept/assertion here represents a domain constructive element or its attribute. In case a domain concept is compound, its elements are grouped by round brackets.

The semantics of a domain concept is defined by semantic tags. Semantic tags are reserved words of NL that are separated by a colon from an NL sentence that contains the names of the domain concepts. In case of necessity the semantics of a domain concept may be specified by additional conditions and restrictions that provide local constraints on the properties of the concepts via the semantic tags **VIZ** and **DEF**.

Depending on the domain situation, the same Reasoning Resource may be a missing, an available or a constructive one. Thus NL provides a semantic polymorphism. In other words, the same reasoning resource has at least three different semantics. It implies that the same name of reasoning resource is associated with at least three meanings. Names of reasoning resources constitute the needs ontology. Every need satisfaction's domain provides a specific ontology mapping. Each need satisfaction's domain provides a specific ontology distribution, assigning to ontological units the semantics defined by the given conditions.

4.2 NL Program as Data

A complete NL program consists of a description of the missing needs, of the available resources and of the constructive resources.

NL program:
Needs
Resources
Activities

The architect of the Need Satisfaction Domain Knowledge Base (knowledge architect) is responsible for the completeness of an NL program. The customer, working with the Need Satisfaction Domain Knowledge Base, must be aware of the origins of his needs. He must know the purpose to meet his needs, and he should be able to describe its properties, major components and resources. However he is not required to have complete information; his information may be uncertain. Both knowledge architect and the customer are not obliged to know the NL language syntax; they are only obliged to answer questions and fill out the templates provided by the system.

1) NL Syntax for Describing a Missing Need

```
NEED: name_of_need
Synonym: (synonyms)
Location: (Why: ...
           For: ...
           Causer: ...)
VIA ListOFneeds: (sub_needs)
Constructive comment:
           (Viz: ... Def: ...)
```

Synonyms help to recognize the semantics of customer specification. The syntax for the need location defines a set of possible triples (Why, For, Causer) that represents the coordinates of a need's semantics that are marked by the semantic tags WHY, FOR and CAUSER:

- WHY: These are the anteceding needs that is a nonempty set of nonempty lists, which contain logical sequences of the anteceding needs
- FOR: These are the ensuing needs that is a nonempty set of nonempty lists which contain logical sequences of the ensuing needs
- CAUSERS: These are the Need causers (optional) that list the subjects or objects whose current state requires a satisfaction of the given need.

ListOFneeds denotes a list of generic sub-needs or disordered set of needs whose satisfaction leads to the given need satisfaction. **Constructive comment** extends the semantics of an NL sentence. It is marked by the semantic tag **Viz**. The attributes of a resource, the conditions of its usage and their previous states, the current state that motivates the given need are published using the semantic tag **Def**.

2) NL Syntax for Describing the Available Resources

NL allows resource descriptions for the following objectives:

1. Description of the generic resources for need satisfaction
2. Description of the generic resources for activity implementation
3. Description of the customer's available resources for need satisfaction

Knowledge architects are responsible for the development of the knowledge related to a certain need and must describe as many variants as possible of the given need satisfaction. Every variant includes a description of a typical resource configuration and a typical activity which satisfies the given need. This implies that for different resources the same need is satisfied by the different ways. Herewith the domain's experts separately describe the resources required for an activity implementation, and activities are carried out in separate domain environments. A customer lists the resources allocated for the given need satisfaction. Optionally he may input the resources required for implementing an activity aimed to his need satisfaction.

The NL Resource syntax includes the semantic tags **BY MEANS OF**, **CEL**, **CE**, **DOC**, **TIME**, **PLACE**, and **BUDGET**.

```
NEED resources:
BY MEANS OF:
CEL: (CE: ...CE: ...)
Doc: ...
Time: ...
Place: ...
Budget: ...
Constructive comment: ...
```

The semantic tag **CEL** is aimed to list constructive resources (enterprises, departments, subdivisions, agents, software agents, equipment, hardware, ingredients, building blocks, ingredients, etc.) that are the subjects or objects of a need satisfaction activity. The semantic tag **CE** is aimed to list the constructive elements that are the subjects or objects of a need satisfaction strategy. The semantic tag **DOC** is aimed to list the instructive elements (contracts, laws, specifications, Charter Company, etc.) that ground the given need satisfaction. The semantic tag **TIME** is aimed to indicate the time resource for the need satisfaction. Semantic tag **Place** is aimed to indicate the resource of place for the need satisfaction. Semantic tag **Budget** is aimed to indicate the scope of financing resource allocated for the need satisfaction.

3) NL Syntax for Describing the Constructive Resources

NL allows description of generic constructive elements (target activities, local activities and objects) and precedents of the activities. A generic activity represents a generalized experience that involves the typical variants of the given need's resources and appropriate activities in the general form. Precedents of an activity represent the known experiences of the generic activity under some known resources and other conditions.

```
Constructive element:
          name_of_ce |
          satisfaction_ behaviour
Synonym: synonyms
Type:
          [target activity |
          local activity |
          object]
```

Location: ...

VIA ListOFneeds: ...

Constructive comment: ...

Need: ...*satisfaction_plan* ...

Precedent of activity: ...

NL allows descriptions of semantic coordinates, marked by the semantic tag Location, not only for missing resources but also for any other NL resource. The semantic coordinates define the locations of an NL resource in the Need Satisfaction Domain knowledge base by a list of semantic tags and their meanings. A satisfaction plan represents a program of sub-needs' satisfaction. The list of needs represents the top operational resource. Every list of needs must be accompanied by an allocation of the need condition resources. A generic activity describes a given need satisfaction by the following steps:

1. definition of the operational resources (i.e., of all the constructive elements and their attributes that meet the given need condition resources in accordance with the initial list of needs)
2. description of the operational list of needs that mirrors the initial list of needs (top operational resources).
3. description of the actions aimed to obtaining the missing data that is necessary for the sub-need satisfaction (In particular, it means the coding of those actions in a programming language or the calling of the corresponding software methods). In other words, in this step the preconditions of the corresponding constructive elements are obtained.
4. description of the actions aimed to processing those preconditions into required preconditions (In particular, it means the coding of those actions in a programming language or the calling of the corresponding software method).
5. listing situations of incompleteness (related to the found postconditions) together with the descriptions of the corresponding ways of the missing data (In particular, it means the coding of those actions in a programming language or the calling of the corresponding software methods).
6. applying steps 3-5 for all the sub-needs as well as for all the constructive elements allocated for every sub-need.

The semantic tag Precedent of an activity marks a record of the typical activity implementation under the concrete conditions and resources. In other words, it represents an experience.

4.3 Example

Suppose a customer needs to make a plan for traveling from Tel-Aviv to Wuhan during Jan 12th and Jan 18th in 2010. In addition, he/she also wants to visit Moscow on Jan 12th, Rostov-on-the-Don on Jan 14th, Beijing on Jan 17th and Shanghai on Jan 18th. He/She also needs to have a day spent in Hon-Kong. The customer picks up a need travel. The query-answering engine generates the following dialogue:

- Which means of transportation do you need?
- Plane.
- What is your point of departure?
- Tel-Aviv.
- How many destinations do you need?
- 6.
- What is the first destination?
- Moscow.
- When you need to arrive in Moscow?
- 12/01/2010
- …
- When you need to arrive in Wuhan?
- 18/01/2010
- What budget do you plan for travel? (Possible answers are: minimum, middle, unlimited).
- Minimum

The query-answering engine translates the customer's answers to the following NL specification:

```
NEED•travel (flight)
BY MEANS OF:
Time:
          from Y(2010).MO(01).D(12)
          to Y(2010).MO(01).D(18)
Place:
          from Tel-Aviv
          to Wuhan
Budget: min
VIA ListOFneeds:
          (
          sub-need: travel(flight)
          BY MEANS OF:
          Time: Y(2010).MO(01).D(12)
          Place: Moscow
          sub-need: travel(flight)
          BY MEANS OF:
          Time: Y(2010).MO(01).D(14)
          Place: Rostov-on-the-Don
```

```
        sub-need: travel(flight)
        BY MEANS OF:
        Time: Y(2010).MO(01).D(17)
        Place: Beijing
        sub-need: travel(flight)
        BY MEANS OF:
        Time: Y(2010).MO(01).D(18)
        Place: Shanghai
        sub-need: travel(flight)
        BY MEANS OF:
        Place: Hon-Kong
        sub-need: travel(flight)
        BY MEANS OF:
        Place: Wuhan
        )
Construct: TSPsolver
call TSPFare ()
```

An example result is shown below:

Table 1. Example of result

Flight NO	**Company**	**Departure**	**Arrival**	**Cost ($)**
Flight 238 Flight 415	AeroSvit Airline	Tel-Aviv	Moscow	326
Flight 367	UT Air	Moscow	Rostov-on-the-Don	86
Flight 602 Flight 910 Flight 111	Aeroflot-Don Air China	Rostov-on-the-Don	Hon-Kong	610
Flight 304	Air China	Hon-Kong	Beijing	362
Flight 333	China Eastern Airlines	Beijing	Shanghai	184
Flight 2508	China Eastern Airlines	Shanghai	Wuhan	143

5 Future Implement

5.1 NL as an Information Retrieval Query Language

As an information retrieval query language, NL is used to help the customer to ask questions (concerning both its service and knowledge base contents) and also to obtain customer's knowledge for satisfying his needs. The semantics of the customer's specifications that the customer describes in form of query/answer is interpreted into NL specifications. The NL search engine interprets the customer's queries into NL specifications and searches the most suitable results. To resolve semantic uncertainty of the results the NL search engine generates clarifying questions, based on the semantic surroundings of the obtained results. After clarifying the sub-needs and available resources the NL search engine looks up in the NL-knowledge base for ready solutions or new assemblage of software on the basis of the solutions found for the sub-needs.

5.2 Activity of the Knowledge Architect

The activities of a knowledge architect are regulated in accordance with the scheme shown in Fig.2

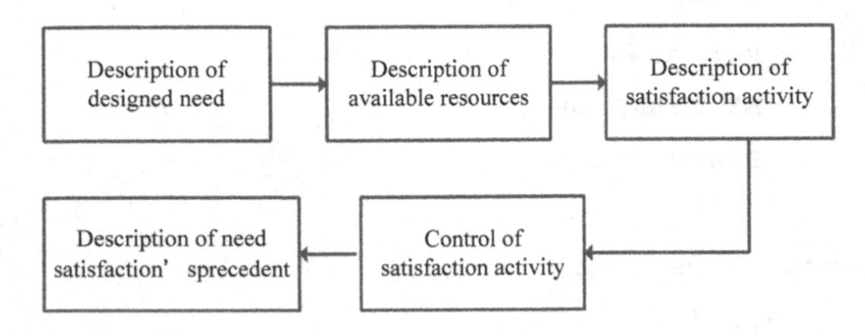

Fig. 2. Knowledge architect activity

A knowledge architect is responsible for carefully filling out NL forms. He is obliged to accurately describe the typical variants of the semantic locations of the needs and all possible combinations of the available resources. Using this information the system engine produces questions to the customer to generate a specification in NL. For any new resource the knowledge architect is obliged to describe its interpretation for all stages of its processing in a multi-domain environment. In other words the knowledge architect grounds the translation of the customer's specification from his professional slang into the terms of API. The knowledge architect describes as many satisfaction activities as possible for every need, either generic or (proved) private.

In describing a generic activity, the knowledge architect provides a sequence of all typical events, connected with shortage of data or with their incorrectness. A description of the satisfaction activity obligatorily includes instructions which control the adequacy of its execution by computers and by other performers. These instructions ground the processing of the intermediate results, asking the questions, answering and an adequate behavior making. The knowledge architect includes in any place of the description the instructions that contain pairs of queries and possible answers as well as a description of the corresponding behavior of the system. Herewith he includes both standard queries and custom ones. The query-answering engine uses these queries and answers and generates additional queries based on the semantic context.

5.3 Satisfying Customer's Needs

The satisfaction of the customer's needs starts from need detection (a customer picks up his need on the so-called Need map) and is supported by the query-answering engine.

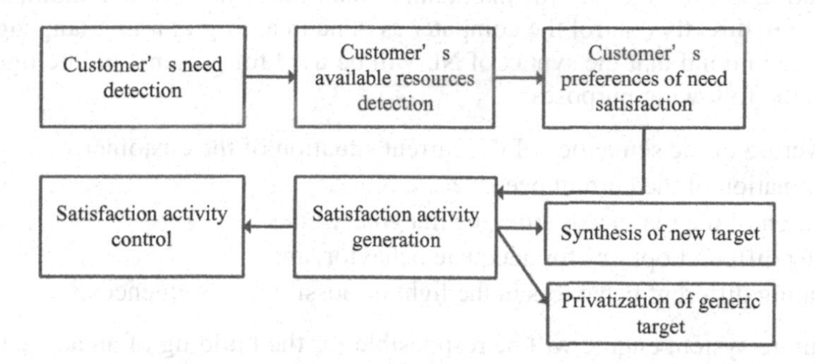

Fig. 3. Customer's need satisfaction

The system engine detects the knowledge contexts which are adequate to a given need and generates the corresponding query-answering interface for discovering the available resources and preferable (for the customer) ways of the given need satisfaction. The system engine is responsible for filling the found satisfaction activity's framework by the available resources (precisely, by properly interpreting the available resources) as well as it is responsible for the reconstruction of this operational framework in accordance with the given semantic context. This procedure is called the Privatization of the generic target activity. In absence of an appropriate target activity the system engine assembles the new satisfaction activity based on the discovered local activities that satisfy the sub-needs of the given need. Finally, the system engine uses the control instructions provided by the knowledge architects for the control of the satisfaction activity.

6 Conclusion

The NL language is a knowledge representation language to describe customers' needs. It has the following advantages:

- To give an opportunity for professionals in various subject areas to represent their experiences in a form that is equally accessible both to the automatic semantic search system and to customers.
- To give an opportunity both for professionals and for every one to use their customized slang to communicate with the computer; in particular, this concerns the description of the current situation and its forerunner.
- To ground a representation of a Need Satisfaction Domain for any private person, for any professional or for the society as a whole.
- To provide an opportunity of building a Global Knowledge Platform that supports different social activities.

- To represent data as a program, and to represent a program as data with the purpose of solving problems using the equation

Decision=Recognition + Synthesis + Control

The NL language is designed more as an intermediate language for capturing and accumulating knowledge and for maintaining man-machine communications, than as a means of to directly control the computer as done in a programming language.

We have in mind that the syntax of NL will be used for guiding an intelligent interface with the following purposes:

- discovering of the semantics of the current situation of the customer,
- determination of the current needs,
- separation of the real needs from the imaginary ones,
- offering different options for adequate behavior, and
- simulating different behaviors in the light of possible consequences.

The future system engine will be responsible for the building of an adequate model of a customer's business (and/or any other) activities. This model will provide prompt reactions to changes of the external conditions.

In addition we can say that the NL language provides along with others the following key needs of the customer:

- need for background information,
- need for training courses,
- need for decision support,
- need for pervasive support of business, educational and/or any other activities.

The result of NL-based processing of the customer's specifications will be a new software product, but more often it will be information that does not require further processing. Thus the NL language will ground the building of software systems of new generation that are capable of managing all resources to satisfy customer's needs.

Acknowledgment

This research has been partially supported by National Science Foundation of China (NSFC) under grant No. 60873007 and the 111 Project of China under grant No. B0737.

References

1. Noy, N., Fergerson, R.W., Musen, M.A.: The knowledge model of Protégé-2000: combining interoperability and flexibility (Retrieved October 11, 2004)
2. Ahsan-ul Murshed, M., Singh, R.: Evaluation And Ranking of Ontology Construction Tools, University of Trento Department of Information and Communication Technology, Technical Report # DIT-05-013
3. Resource Description Framework, http://www.w3.org/RDF/
4. OWL Web Ontology Language Overview, W3C Recommendation (February 10, 2004), http://www.w3.org/TR/owl-features/

5. Lloyd, J.W.: Practical Advantages of Declarative Programming. In: Proceedings of the 1994 Joint Conference on Declarative Programming (1994)
6. Kowalski, R.A.: Algorithm = Logic + Control. CACM 22(7) (1979)
7. Turner, D.A.: Programs as executable specifications. In: Hoare, C.A.R., Shepherdson, J.C. (eds.) Mathematical Logic and Programming Languages (1985)
8. McCarthy, J.: Programs With Common Sense. Computer Science Department. Stanford University, Stanford, CA 94305
9. Abramovich, A.: Domain knowledge representation, SKLSE, Wuhan University, China
10. Evans, E.: Domain-Driven Design - Tackling Complexity in the Heart of Software. Addison-Wesley, Reading (2004)
11. Sowa, J.F.: Semantic Networks, Encyclopedia of Artificial Intelligence (1987) (Retrieved April 29, 2008)
12. Hjørland, B., Albrechtsen, H.: Toward A New Horizon in Information Science: Domain Analysis. Journal of the American Society for Information Science, 1995 46(6), 400–425 (1995)
13. Petras, V.: Translating Dialects in Search: Mapping between Specialized Languages of Discourse and Documentary Languages. University of California, Berkeley (2006) (Dissertation), `http://www.ischool.berkeley.edu/~vivienp/diss/vpetras-dissertation2006-official.pdf`
14. RDF Semantics, W3C Recommendation (February 10, 2004), `http://www.w3.org/TR/rdf-mt/`
15. Pinkal, M., Koller, A.: Semantic Theory, Lexical Semantics (Summer 2005), `http://www.coli.uni-saarland.de/courses/semantics-05/lectures/lect13.pdf`
16. Bunge, M.: Causality: The Place of the Causal Principle in Modern Science. Harvard University Press, Cambridge (1959)
17. Iscoe, N.: Domain modeling-overview &Ongoing research at EDS. In: Proceedings of the 15th international conference on Software Engineering, Baltimore, Maryland, United States, pp. 198–200
18. Aamodt, A.: A knowledge representation system for integration of general and case-specific knowledge, Tools with Artificial Intelligence. In: Proceedings of Sixth International Conference, New Orleans, LA, USA, November 6-9, pp. 836–839 (1994)
19. Bahrami, M., Kaviani, S.: A New Method for Knowledge Representation in Expert System's (XMLKR). In: First International Conference on Emerging Trends in Engineering and Technology, ICETET 2008, July 16-18, pp. 326–331 (2008)
20. Sun, Y., Zhiping, L.: Ontology-based domain knowledge representation. In: Processings of 2009 4th International Conference on Computer Science & Education, pp. 174–177 (2009)
21. Tian, F., LiRen, Y., Kuroiwa, F.: Ontology based domain knowledge construction. In: Natural Language Processing and Knowledge Engineering, August 30-September 1, pp. 176–182 (2007)
22. Greiner, R., Lenat, D.: A representation language lanuage. In: Proceeding AAAI 1980, pp. 165–169. Morgan Kaufmann, San Francisco (1980)
23. Wilensky, R.: Knowledge respresentation, a critique and a proposal. In: Kolodener, J.L., Riesbeack, C.K. (eds.) Experience, memory and resoning, pp. 12–28. Lawrence Erlbaum, Mahwah (1986)

Adding Taxonomies Obtained by Content Clustering to Semantic Social Network Analysis

Hauke Fuehres[1], Kai Fischbach[2], Peter A. Gloor[1], Jonas Krauss[1], and Stefan Nann[1]

[1] Center for Collective Intelligence MIT, Cambridge, MA, USA
[2] Department of Information Systems and Information Management, University of Cologne

Abstract. This paper introduces a novel method to analyze the content of communication in social networks. Content clustering methods are used to extract a taxonomy of concepts from each analyzed communication archive. Those taxonomies are hierarchical categorizations of the concepts discussed in the analyzed communication archives. Concepts are based on terms extracted from the communication's content. The resulting taxonomy provides insights into the communication not possible through conventional social network analysis.

1 Introduction

People increasingly publish information on their social networks on the Internet and on social network sites in special [1]. Various sources can be used to obtain the data on relations between different actors. Email archives as well as publicly available online-forums may serve, among others, as the sources of data to be analyzed [2]. Those interaction networks can be studied through measures of *social network analysis (SNA)*. Analyzing social networks with these measures reveals the structure of the networks. The information needed to model the network is often explicitly given or can easily be obtained.

The aim of this work is to combine the analysis of (electronically available) communication structures by means of social network analysis with the analysis of communication content by means of *information retrieval (IR)* and to introduce a software tool performing these tasks. This tool has been implemented as a module of the Condor toolkit. The content of communication sent in social networks is analyzed using information retrieval. The topics discussed in those communication messages are extracted and visualized. The automated construction of a taxonomy from the extracted topics helps to understand the relationships between them.

In contrast to the data representing the network structure, the content of the communication is usually unstructured. Interactions in the surveyed networks are often unstructured documents sent from one actor to one or more other actors, sometimes enriched with additional attributes like a timestamp. IR methods help to analyze the unstructured message content. Essential to those methods, like

T.J. Bastiaens, U. Baumöl, and B.J. Krämer (Eds.): On Collective Intelligence, AISC 76, pp. 135–146.
springerlink.com

flat- or hierarchical clustering, is the introduction of a similarity measure on the words used in the communication content.

The idea behind the proposed approach is twofold: On one hand, the formal analysis of the communication structure by SNA methods is enriched with information on the communication content. Supplementary to the key insights gained by SNA methods, statements on the topics discussed by certain actors can be made. On the other hand, satisfying information needs using information retrieval can be supported by the formal knowledge on the network structure. SNA offers methods to assess the centrality of each actor in a network. Those key figures of actors are used to weight the information retrieved from messages they sent. Extraction of topics from the messages should not only be based on the topics' importance in the "flat text-file" but also show the context they are used in. A person's information acquisition is influenced by their social network [3]. Weighting in key data on those networks into the information retrieval process reflects the importance of those people's role in the information acquisition and diffusion process.

2 Related Work

Although the role of social network structures in document corpora is well known and utilized for information retrieval tasks, surprisingly little work deals with the extraction of term relations from the content of social networks. Usually the social network structure of documents is used to improve the document weights. Those weights modify the sorting order of retrieved documents in a search engine. A typical example of such a social network structure to be utilized to improve document ranking is the hypertext structure of web documents [4].

When focussing on the extraction of terms and leaving out the SNA component different information retrieval approaches can be identified. Supervised learning methods like the support vector machine are a prominent example [5]. This classifying technique was adapted to diverse purposes and requirements. One enhancement to the SVM relevant to the scope of this paper is the hierarchical support vector machine proposed in [6]. Instead of constructing a "flat classification", this technique allows to create hierarchical taxonomies.

Another important way of extracting topics from text corpora can be achieved by utilizing *latent semantic indexing (LSI)* [7]. It is an enhancement to the vector space model described in [8]. LSI is based on a singular value decomposition (SVD) of the term-document matrix. After calculating the SVD, the term-document matrix can be approximated with a lower-rank matrix. With this step a dimension reduction of the term-document vector space can be achieved. Instead of using the term-document vector space, IR methods can work on the reduced concept space defined by the SVD. The clustering algorithms applied to the problem of finding taxonomies of terms can work on the reduced concept space instead of using the original vector space. But more important to the goals of this paper is the ability to reveal hidden structures in the original vector space. Those hidden structures to be uncovered are polysemes and synonyms.

Having access to this information on terms may help to improve the quality of taxonomies obtained by clustering algorithms. Preprocessing with a stemmer can also be avoided since LSI will identify words with similar meanings as related concepts.

3 Semantic Social Network Analysis

SNA provides methods to analyze the interactions and relationships between actors in a network. The field of SNA emerged in behavioral science where interactions of people were analyzed. At the beginning there was the insight that interactions between individuals have influence on the individuals themselves [9, ch. 2]. Analyzing different relations between actors has been applied to different fields of study [10]. The methods have been successfully applied to Organisational Behavior [11] and the analysis of the spread of diseases [12].

Semantic social network analysis factors in content analysis of the relational data into the analysis of social networks. Therefore, it can be applied to social network data where interaction takes place by exchanging textual information. These can be found in email archives and networks built of websites and their linkage among each other. Another prominent example of textual interaction shared in a social network are the messages exchanged in online forums.

Semantic social network analysis as introduced in [13] allows to analyze the content of textual interaction in social networks together with the network structure. Techniques from information retrieval are used to extract important terms from the interaction's content.

Analyzing networks can be conducted in a static or a dynamic way. Traditional SNA focuses on a static view on the available data. The key measures used in SNA reflect a social network in a static way. However, networks analyzed with the means of SNA can be of dynamic nature. Networks might evolve over time. Identifying and understanding patterns in an evolving network can help to understand the nature of the whole network [14]. One possibility to gain insight into the dynamic structure of a network is to divide the data on the social network into several timeframes. The next step is to calculate SNA key measures for each of those timeframes and compare them over time [15].

4 Our Approach: Clustering

The goal of this work is to extract a *taxonomy* of terms and concepts from the interaction's content. This taxonomy should give an overview of the discussed topics in the content of the interaction in the analyzed social network. The elements to be categorized are the terms extracted from the interaction's content. A taxonomy created on top of those terms should help to understand the most important topics discussed in the interactions between actors of the analyzed social networks. Different ways of obtaining a (hierarchical) classification of a set of discriminable objects are known. In this section clustering methods are introduced. Generally, these methods are used to assign discriminable objects into groups. Clustering methods are applied in different fields of research. Those methods are used to group unlabeled data. Although the underlying idea of clustering methods is identical in all those fields, various terminologies and assumptions emerged [16]. In information retrieval clustering is often used to locate information. Using clustering to find information is used for a long time in libraries where books are classified by the topic they discuss [17].

Besides calculating the similarity of terms by utilizing their distribution among the analyzed documents, the similarity of terms can also be obtained from external sources. The idea behind this approach is to measure the similarity of terms in the analyzed documents by obtaining the semantic similarity of those terms from an existing taxonomy. Those taxonomies, or generally speaking lexical networks, can be obtained from different kinds of sources. The way on how to calculate the semantic similarity between two words might differ depending on the source of the lexical network used [18]. However, the similarity measures can be categorized into two different approaches [19]. The first category combines edge counting based methods whereas the second category roots in information theory based methods. Both approaches are described later on in this section.

The base of the similarity measures described in this section are lexical networks or lexical taxonomies. A lexical taxonomy is a tree-like structure with its nodes representing concepts. One source of background knowledge for measuring semantic similarity of words is the Wikipedia online encyclopedia. Wikipedia is an encyclopedia built on user generated content. It has a general scope with more than 3,222,261 articles in the English Wikipedia.[1] In Wikipedia authors are encouraged to add existing or new pages to categories and create new categories when necessary. The categories are arranged in a category network with a tree-like structure. This network of categories can be used to derive a semantic taxonomy. Several ways of extracting semantic taxonomies from Wikipedia are known. In [20] Wikipedia categories are used to identify topics of documents by relating the documents' content to category titles and to the titles of articles in categories. In [21] the structure of the Wikipedia's category network as well as the titles of the categories are used to extract semantic relations between different concepts.

4.1 Edge Counting Based Similarity

The first family of semantic similarity measures are edge counting based measures. Those similarity measures use the number of edges between two concepts in the graph representing the semantic network to calculate the similarity of those concepts. In [19] the basis of the edge counting measures is seen in [22]. A simple approach to calculate the similarity of two concepts is to use the path length of the shortest path from one concept to the other as the measure of similarity [23].

[1] Number from http://en.wikipedia.org/wiki/Special:Statistics in March 2010.

4.2 Information Based Similarity

Instead of using the edge counting similarity measures described in the previous section, an information based approach is introduced in [23] to calculate the distance between different semantic concepts. How much information two concepts share in common is the intuition of this similarity measure. The idea behind this approach is based on the information theoretical notion of information content of a concept. For each concept c in the taxonomy, $p(c)$ is the probability of encountering an instance of concept c. The higher a concept c_i is placed in the taxonomic tree the higher is its probability $p(c_i)$. If the taxonomy has one single root node its probability is 1. The *information content* of a concept c is $ic(c) = -\log p(c)$ and serves as the foundation for the calculation of the information based similarity. This information content decreases with the increasing of its probability. That means the more general a concept is the lower is its information content. The concept embodied by a single root node in a semantic taxonomy therefore has an information content of 0. In order to gain specific values for the probabilities of the concepts the frequencies of words in natural language corpuses can be used. The similarity of concepts based on the information content is defined by:

$$sim_{RES}(c_i, c_j) = \max_{c \in S(c_i, c_j)} -\log p(c) \tag{1}$$

with $S(c_i, c_j)$ being the set of concepts subsuming both c_i and c_j[23]. With this definition the similarity of two concepts c_i and c_j in a semantic taxonomy is measured by the information content of the lowest common subsumer (LCS) of c_i and c_j. The lowest common subsumer of c_i and c_j is the lowest node in the semantic taxonomy that subsumes concepts c_i and c_j and thus is a hypernym of both concepts.

A notable generalization of information based similarity measures was introduced in [24]. The aim was a universal and theoretically justified model of similarity. Whereas other measures are bound to a particular domain or application, Lin's measure is only based on information theory. This omits assumptions based on the underlying domain. The definition of Lin's similarity is rooted in assumptions on the concept of similarity not in any specific formula. Different similarity measures for specific domains can be derived from those assumption. The derived semantic similarity is similar to Resnik's similarity measure:

$$sim_{LIN}(c_i, c_j) = \frac{2 * \log\ p(LCS(c_i, c_j))}{\log p(c_i) + \log p(c_j)} \tag{2}$$

with $LCS(c_i, c_j)$ defining the lowest common subsumer of c_i and c_j.

4.3 Boosting Similarity of Terms with the betweenness Centrality of the Actors

Although the corpuses analyzed with the introduced implementation can be of different nature they all have a social network structure in common. Social network analysis provides information on the structure of a network and on the position of actors in such a network. One powerful tool to assess the importance of an actor in a social network is the betweenness centrality of each actor. This measure reveals an actor's degree of centrality in a social network. Centrality thus can be interpreted as the importance of an actor. A message send by an actor can be linked to the importance of the sending actor. Thus a message can be weighted with the centrality of its sending actor.

Messages of less important actors can now be identified. In this way it possible to take only messages of important actors into account. The basis of analyzing terms and their similarity in this work is built on the vector space model. In a term-document vector space the distinction between different levels of importance of documents can be used for pruning the dimensionality of the vector space. Documents with a low importance can be ignored in the following analysis.

Besides reducing the dimensionality of the term-document vector space the importance weights of the documents can be used to calculate an importance weight for each term. Such an importance weight for a term is based on the importance weights of the documents and therefore is based on the betweenness centrality of the actors in the social network.

The hierarchical methods can work either by pooling all objects into one cluster and splitting up this cluster recursively or by starting with single objects and merging them into clusters. *Top-down clustering*, although less frequently used, has some advantages over merging algorithms. It is possible to stop the calculation when the clusters are fine-grained enough. Also, the global distribution of objects to cluster is taken into account [25, p. 396]. When using bottom-up or *hierarchical agglomerative clustering (HAC)* algorithms all decisions in the clustering process are made on a local basis without taking the global distribution into account. On the other hand, the top-down algorithms are more complex since flat clustering techniques are applied for each cluster to be split. In each step of an HAC algorithm the most similar clusters are merged. This procedure iterates till only one cluster is remaining that holds all terms. Alternatively the algorithm might be designed to stop when a certain number of top-level clusters remain. The more common hierarchical agglomerative clustering algorithms are discussed in the following sections and are the foundation of the described implementation.

Complete-Link Clustering. The clustering algorithms depend on similarity measures on the terms to cluster. The introduced similarity measures are defined as functions $sim : \mathbb{T} \times \mathbb{T} \rightarrow [0, 1]$ with $\mathbb{T}$ being the space of terms to be compared. Since these functions are defined on the binary Cartesian product of the term vector space, new similarity measures are needed for comparing similarity of clusters of terms. Such a similarity measure of clusters needs to compare more than two terms with each other. These similarity measures on clusters yield the clusters to merge in each step by determining the most similar clusters.

A simple way of calculating the similarity of two clusters is *single-link clustering*. In each step, this algorithm merges the clusters with the nearest neighboring members. The single-link clustering is a local criterion since only one singleton member of each cluster is relevant to the calculation of the similarity. The similarity of the most similar members is the similarity of both clusters. The single-link clustering was introduced in [26]. Another way of calculating the similarity of clusters is the *complete-link clustering* algorithm. Instead of only paying attention to the most similar members of two clusters to calculate the cluster similarity, the diameter of the merged cluster is the crucial measure. The similarity of two clusters is the diameter of the merged cluster. This similarity can be calculated by assigning the similarity of the two most dissimilar singletons. In contrast to the single-link clustering the complete-link clustering is not local since the diameter of the whole cluster is taken into account. Therefore, the resulting clusters are more compact; clusters with small diameters are preferred by this method.

Group-Average Agglomerative Clustering. The complete-link clustering introduced in the previous section chooses only one representing member of each cluster to calculate the similarity with the other clusters. Even though the diameter of each cluster is taken into account by the complete-link clustering, obstacles like the sensitivity against outliers persist. The *group-average agglomerative clustering (GAAC)* method uses the similarity of each member of the clusters to calculate the similarity of clusters. The aim of the GAAC algorithm is to build compact clusters. The average of the pairwise similarities of all members of both clusters is calculated [25]. It is important to mention that the similarity of members already in the same cluster is also taken into account.

By using the GAAC algorithm, the behavior of single-link clustering algorithms to create chains of clusters is avoided as well as the strong sensitivity towards outliers of complete-link clustering.

Since each member of each cluster is factored in in each step of calculating similarities and merging the most similar clusters, the time complexity can not be reduced with priority queues, as it is possible for single-link and complete-link algorithms. Thus the time complexity of the GAAC algorithm is in $\mathcal{O}(N^3)$. [27] shows how the complexity can be reduced to $\mathcal{O}(N^2)$, although with several constraints. This simplification only holds when the objects to cluster are represented by vectors in $\mathbb{R}^N$ and the applied similarity measure is the dot product. The key to the simpler calculation of the similarities of cluster C_i and cluster C_j is the definition of cluster similarity by: [25, p. 389]

$$sim_{GAAC}(C_i, C_j) = Norm(C_i, C_j) \left[(\sum_{t_m \in C_i \cup C_j} t_m)^2 - (|C_i| + |C_j|) \right] \quad (3)$$

with:

$$Norm(C_i, C_j) = \frac{1}{(|C_i| + |C_j|)(|C_i| + |C_j| - 1)} \quad (4)$$

5 Implementation and Evaluation

In this implementation the semantic social network analysis [13] module of Condor, the successor of TecFlow [28], is extended. Condor analyzes and visualizes communication and interaction networks. The data representing the network structure needs to be available in electronic format. Condor natively supports different data sources. Email archives can be imported from Eudora, Microsoft Outlook, or directly from an IMAP server. Weblinks and blogs can be accessed via Google's blogsearch and Microsoft's live search API. Data gathered with Social Badges can also be loaded into Condor [29]. Not natively supported data sources can be loaded into Condor via flat files or by parsing the data directly into a MySQL database. In extension to a static analysis of interaction networks, Condor and its predecessors can be used to study dynamic networks and their evolution over time. Monitoring social networks over time helps understanding the evolution of relationships in the networks [15]. Condor visualizes the social networks with a spring-embedder model developed by Fruchterman and Reingold [30]. This algorithm enhances Eades method [31] to places the nodes and the edges of a network on a two-dimensional plane. The screenshot in figure 3 shows the resulting structure of the force-directed algorithm.

An important feature of Condor is its ability to process and visualize temporal information on social networks. Especially when analyzing the content of the communication the temporal distribution of terms used by actors in the network is of interest. An increased use of terms pooled in certain clusters at one moment in time could point to the topics discussed in the social network during that time. To support users in assessing the prominence of single nodes in the taxonomies and the concepts these nodes represent this implementation allows to view the temporal distribution of each node in a chart. Figure 1 shows an example of such a distribution chart.

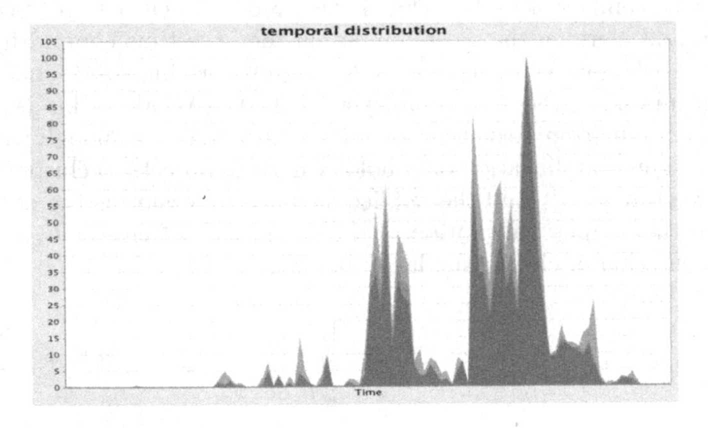

Fig. 1. Screenshot of the temporal distribtion chart

In order to contrast the cluster scrutinized by the user to all the other clusters in the taxonomy the distribution of all terms is shown in each chart too. Values used in the temporal distribution chart are weighted with the terms' importance scores as described in section 4.3. By using those importance scores instead of the bare numbers of term occurrences, the betweenness centrality of the actors using the terms is factored in. Utilizing the betweenness based importance of each term punishes those terms and concepts used by less important actors.

An additional view on the data in each cluster is the list of the most important documents shown in figure 2. For each document a summary as well as the sender's name and the submission date is given. The documents are ordered by their importance for the terms in the selected cluster. On the left side of figure 2 an additional window with the content of the selected document can be seen.

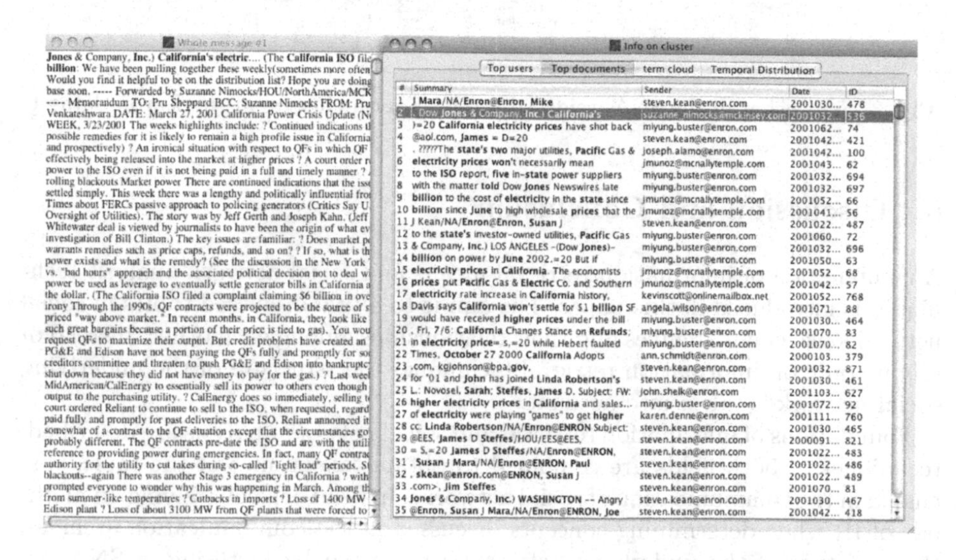

Fig. 2. Screenshot of the list with the most-important documents

The evaluation of clustering algorithms can be conducted with statistical measures as introduced in [32]. Those methods assess the quality of the clustering algorithms. However, the quality of the resulting clustering needs to be judged by human users. A reduced Enron dataset[2] is used to show the functionality of the developed module. This dataset consist of emails collected from Enron employees during the Enron scandal. Figure 3 shows the top-level clusters of this dataset. Important facts on the Enron scandal can be grasped at a glance without further knowledge on the background of this dataset. The system identifies Enron's Executive Vice President Steven J. *Kean* as one of the key players in the scandal in which *attorneys* and *New York* played a crucial role.

[2] Available at `http://www.cs.cmu.edu/~enron/`.

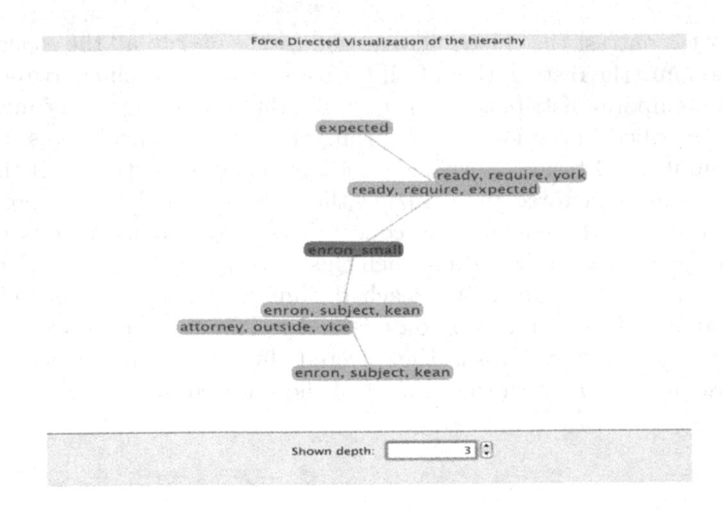

Fig. 3. Screenshot of the SNA-weighted top-level term clusters in the Enron dataset

6 Discussion and Conclusion

A combination of information retrieval means and social network analysis techniques is introduced in this paper. The aim is to reveal discussed topics in social networks like email archives and their relationships among each other. Instead of relying only on information retrieval techniques the structure of the underlying social network is taken into account.

Foundations of information retrieval and social network analysis are described, techniques of both fields are combined to obtain taxonomies of the topics discussed in the communication of social networks. Instead of solely relying on methods of IR when determining concepts discussed in the communication archives, the structure of the underlying network is respected by factoring in SNA key measures.

Unsupervised clustering algorithms are used to scrutinize the content of communication in social networks. Classification of terms with those algorithms is a new approach to gain insights on communication networks. The resulting taxonomies of terms can be used to obtain an overview on the whole communication network at a glance. A temporal analysis module allows assessing the development of discussed topics over time. Finally, the design and details of the implementation are presented.

The aim of this paper was to provide users of social network analysis packages like Condor with an automated method to reveal topics discussed in analyzed networks and their hidden relations among each other. First steps were made in this paper to allow users to gain an impression on the nature of discussions in analyzed networks. This work can only serve as a step in the right direction of automatically revealing discussed topics in social networks. Supporting users with enhanced automated or semi-automated methods to analyze the content of social networks is crucial with more and more data available on social networks.

References

1. Boyd, D., Ellison, N.B.: Social network sites: Definition, history, and scholarship. Journal of Computer-Mediated Communication 13(1-2) (November 2007)
2. Gloor, P.A., Cooper, S.M.: Coolhunting: chasing down the next big thing. Amacom, New York (2007)
3. Borgatti, S.P., Foster, P.C.: The network paradigm in organizational research: A review and typology. Journal of Management 29(6), 991–1013 (2003)
4. Korfiatis, N., Sicilia, M.A., Hess, C., Stein, K., Schlieder, C.: VI. In: Social Network Models for Enhancing Reference Based Search Engine Rankings, pp. 109–133. Idea Group Reference (2007)
5. Vapnik, V.N.: The Nature of Statistical Learning Theory. Springer, New York (2000)
6. Cai, L., Hofmann, T.: Hierarchical document categorization with support vector machines. In: CIKM 2004: Proceedings of the thirteenth ACM international conference on Information and knowledge management, pp. 78–87. ACM, New York (2004)
7. Deerwester, S.C., Dumais, S.T., Landauer, T.K., Furnas, G.W., Harshman, R.A.: Indexing by latent semantic analysis. Journal of the American Society of Information Science 41(6), 391–407 (1990)
8. Salton, G., Wong, A., Yang, C.S.: A vector space model for automatic indexing. Communications of the ACM 18(11), 613–620 (1975)
9. Scott, J.P.: Social Network Analysis: A Handbook. SAGE Publications, Thousand Oaks (2000)
10. Freeman, L.C.: Social Network Analysis: Definition and History. Encyclopedia of Psychology 6, 350–351 (2000)
11. Krackhardt, D., Brass, D.: Intra-Organizational Networks: The Micro Side, pp. 209–230. Sage Publications, Thousand Oaks (1994)
12. Klovdahl, A.S.: Social network research and human subjects protection: Towards more effective infectious disease control. Social Networks 27(2), 119–137 (2005)
13. Gloor, P.A., Zhao, Y.: Analyzing actors and their discussion topics by semantic social network analysis. In: IV 2006: Proceedings of the conference on Information Visualization, pp. 130–135. IEEE Computer Society, Washington (2006)
14. Carley, K.M.: Dynamic network analysis. In: Breiger, K.C.R., Pattison, P. (eds.) Dynamic Social Network Modeling and Analysis: Workshop Summary and Papers. Committee on Human Factors, National Research Council, pp. 361–370 (2003)
15. Gloor, P.A.: Capturing team dynamics through temporal social surfaces. In: IV 2005: Proceedings of the Ninth International Conference on Information Visualisation, pp. 939–944. IEEE Computer Society, Washington (2005)
16. Jain, A.K., Murty, M.N., Flynn, P.J.: Data clustering: a review. ACM Comput. Surv. 31(3), 264–323 (1999)
17. Kowalski, G., Maybury, M.T.: Information Storage and Retrieval Systems: Theory and Implementation. Kluwer Academic Publishers, Norwell (2000)
18. Budanitsky, A., Hirst, G.: Evaluating wordnet-based measures of semantic distance. Computational Linguistics 32(1), 13–47 (2006)

19. Li, Y., Bandar, Z., McLean, D.: An approach for measuring semantic similarity between words using multiple information sources. IEEE Transactions on Knowledge and Data Engineering 15(4), 871–882 (2003)
20. Schonhofen, P.: Identifying document topics using the wikipedia category network. In: WI 2006: Proceedings of the 2006 IEEE/WIC/ACM International Conference on Web Intelligence, pp. 456–462. IEEE Computer Society, Washington (2006)
21. Nastase, V., Strube, M.: Decoding wikipedia categories for knowledge acquisition. In: Fox, D., Gomes, C.P. (eds.) AAAI, pp. 1219–1224. AAAI Press, Menlo Park (2008)
22. Rada, R., Mili, H., Bicknell, E., Blettner, M.: Development and application of a metric on semantic nets. IEEE Transactions on Systems, Man and Cybernetics 19(1), 17–30 (1989)
23. Resnik, P.: Using information content to evaluate semantic similarity in a taxonomy. In: International Joint Conference for Artificial Intelligence (IJCAI 1995), pp. 448–453 (1995)
24. Lin, D.: An information-theoretic definition of similarity. In: ICML 1998: Proceedings of the Fifteenth International Conference on Machine Learning, pp. 296–304. Morgan Kaufmann Publishers Inc., San Francisco (1998)
25. Manning, C.D., Raghavan, P., Schütze, H.: Introduction to Information Retrieval. Cambridge University Press, Cambridge (2008)
26. Sneath, P.H.A., Sokal, R.R.: Numerical taxonomy: the principles and practice of numerical classification. Freeman, San Francisco (1973)
27. Cutting, D.R., Karger, D.R., Pedersen, J.O., Tukey, J.W.: Scatter/gather: a cluster-based approach to browsing large document collections. In: SIGIR 1992: Proceedings of the 15th annual international ACM SIGIR conference on Research and development in information retrieval, pp. 318–329. ACM, New York (1992)
28. Gloor, P.A., Zhao, Y.: Tecflow - a temporal communication flow visualizer for social networks analysis. In: CSCW 2004 Workshop on Social Networks. ACM, New York (2004)
29. Gloor, P.A., Oster, D., Putzke, J., Fischbach, K., Schoder, D., Ara, K., Kim, T., Laubacher, R., Mohan, A., Olguin Olguin, D., Pentland, A., Waber, B.N.: Studying microscopic peer-to-peer communication patterns. In: Proceedings AMCIS Americas Conference on Information Systems (2007)
30. Fruchterman, T.M.J., Reingold, E.M.: Graph drawing by force-directed placement. Softw. Pract. Exper. 21(11), 1129–1164 (1991)
31. Eades, P.A.: A heuristic for graph drawing. In: Congressus Numerantium, vol. 42, pp. 149–160 (1984)
32. Stein, B., zu Eissen, S.M., Wißbrock, F.: On cluster validity and the information need of users. In: Proceedings of the Artificial Intelligence and Applications Conference (2003)

How to Reduce New Product Development: Customer Integration in the e-Fashion Market

Frank T. Piller and Evalotte Lindgens

TIM Research Group, RWTH Aachen University, Germany
piller@tim.rwth-aachen.de, lindgens@tim.rwth-aachen.de

Abstract. Forecasting the demand for new products is becoming increasingly difficult in many markets. A new method to decrease the flop rate of new products is the idea to integrate customers deeply into the innovation process. This method of integrating the commitment of users to screen, evaluate and score new designs as a powerful mechanism to reduce flops of new products. The process starts when an idea for a product is posted on a dedicated web site by either a (potential) customer or just the designer of a product. Second, reactions and evaluations of other consumers towards the posted idea are encouraged in form of internet forums and opinion polls. Based on the results of this process, the manufacturer investigates the possibility of commercialization of the most popular designs. Is this evaluation positive, the company decides about a minimum amount of purchasers necessary to produce the item for a given sales price, covering its initial development and manufacturing costs (and the desired margin). The new product idea is then presented to the customer community, and interested customers are invited to express their commitment to this idea by voting for the design or even placing an order. Accordingly, only if the number of interested purchasers exceeds the minimum necessary lot size, investments in final product development are made, merchandising is settled and sales are commenced.

1 Introduction

The manufacturer's nirvana is to develop and produce exactly *what* its customers want and *when* they want it – ideally with no risk of overstocks or inventory. The increasing heterogeneity of demand, a rapid change of preferences, and the resulting micro-segmentation of many product categories however prevents manufacturers to reach this state easily. In many consumer goods markets, manufacturers today are forced to create fitting assortments for smaller market niches than ever, as these markets frequently are the only way for growth and to escape from heavy price competition. In such a situation, new product development projects often cause enormous investments and are highly risky. While new products or product variants have to be developed and introduced at high pace, forecasting their exact specification and potential sales volumes is becoming more difficult than ever. Recent research studies confirm large failure rates in new product commercialization.[1] Newly launched products have shown notoriously high failure rates over the years, often reaching fifty

T.J. Bastiaens, U. Baumöl, and B.J. Krämer (Eds.): On Collective Intelligence, AISC 76, pp. 147–158.
springerlink.com

percent or more. The primary reason for these flops has been found to be inaccurate understanding of user needs. Many new product development projects are unsuccessful because of poor commercial prospects rather than due to technical problems. Research found that timely and reliable information on customer preferences and requirements is the most critical information for successful product development.[2] Conventionally, heavy investments in market research are seen as the only measure to access this information. Especially the Apparel Industry with its fast changing trends and collections, where companies like H&M get new designed clothes every 3 weeks and therefore the assortment no longer only changes four times a year, is faced with huge challenges. How will one identify perfectly the customers' needs to forecast their future desires and design and produce on this basis optimal fitting apparel? One opportunity to handle these challenges is shown by an extraordinary company called Threadless. Besides reducing inventories, eliminating of markdowns and increasing customer loyalty, they do a marvelous thing: producing exactly what the customer wants – by asking him. So far it does not seem like there is any difference to a common company- most of them "ask" their customers "what they want" by market research. The clue on Threadless' "asking the customer" is, that they ask to score every single product that is online, moreover to decide if the customer would buy it- and after all please him to change the product itself to more fit the customer's needs.

Contrast Threadless' model of collecting customer purchase orders in advance of expenditures on detailed design and production with the conventional model of conducting market research and building agile manufacturing systems. Common wisdom says that to learn about customer preferences and requirements, companies should invest in market research activities. To transfer this information into fitting assortments with short lead times, many companies have built large systems of quick response manufacturing or even mass customization. But these measures are often costly and do not deliver what companies expect.

Consider market research: Questionnaires, surveys, or interviews ask consumers what they like and dislike. Among the methods for testing new concepts, the most common are focus groups. They are popular because the results are easy to interpret and the method is fast, inexpensive, flexible, and confidential. Unfortunately, focus group research has a number of severe limitations.[3] One problem is that the results from a test with a few consumers are not a reliable indicator of the reactions of the broader population. In addition, focus groups lack realism. Consumers have to react to verbal descriptions of concepts or a rendering of a product. As a consequence, this research method tends to underestimate the benefits of a truly unique new product concept. Focus group research – and most other common market research methods – also does not measure real consumer purchasing behavior. It reveals information about the consumers' attitudes toward new products or their intentions to purchase them. But it does not provide quantitative estimates of sales, market share, product cannibalization, and profitability. More reliable and accurate measures like test markets are demanding expensive set-ups and take very long to deliver results. Also, there is a high level of noise in these tests like competitors' activities, manufacturers' advertising, and economic change. Finally, most market research measures demand background data to calibrate forecasting or to correct for biases in stated purchase intentions. This data may be available in established categories for consumer packaged-goods, but not for radical innovations or products targeting highly heterogeneous market segments.

Anticipating these problems, many companies perform no market research at all. Studies of the actual practice of market research report that companies regularly fail to undertake thorough market research and use only very few of the available tools and methods to include customer input in the development process. A survey of Fortune 500 firms found that only the focus groups method was used by more than the half of the companies studied, and only two other methods (limited rollouts and concept tests) were used by more than 25% of the respondents.[4] This is rather surprising, given the huge amount of scholarly study and a whole industry providing these market research services. One frequent excuse is that customers are difficult to predict: they often cannot express what they want or are internally inconsistent, often many people with different needs are involved in one purchase decision, and it is likely that customers have changed their mind by the time the product is launched. As a result, many manufacturers tend to stick with existing assortments, building their new products first of all on a revision of the existing offerings. This may improve the capability to forecast demand for new variants, but places suppliers in a persistent danger to miss important trends. It also prevents them to surprise their customers with really new products and innovative applications.

2 Threadless.com's Idea to Substitute Market Research Expenditures by Sales

But Threadless, a young Chicago-based fashion company follows an innovative business model that allows it to create a high variety of products without risk and without heavy investments in market research to access customer preferences before production starts. In fact, it follows a strategy that turns market research expenditures into quick sales. Started in 2000 by designers Jake Nickell and Jacob DeHart, Threadless focuses on a hot fashion item, t-shirts with colorful graphics. This is a typical hit-or-miss product. Its success is defined by fast changing trends, peer recognition, and finding the right distribution outlets for specific designs. Despite these challenges, none of the company's many product variants ever flopped. But Threadless has neither sophisticated market research or forecasting capabilities nor a complicated flexible manufacturing system.

Rather, all products sold by Threadless are inspected and approved by user consensus before any larger investment is made into a new product. Only after a sufficient number of customers have expressed their explicit willingness to buy a new design, the garment is produced. If this commitment is missing, a potential design concept is dismissed. But if enough customers pledge to purchase the product, the design will be finalized and go into production. In this way, market research expenditures are turned into early sales. New designs regularly sell out fast, but are reproduced only if a large enough number of additional customers commit to purchase a reprint. Some customers are even integrated deeper in the new product development process. All new designs are submitted entirely by the community, which includes hobbyists, but also professional graphic designers. The company exploits a large pool of talent and ideas to get new designs (much larger than it could afford if the design process would have been internalized). Creators of submissions which are selected by other users get a $2000 reward, and their name is printed on the particular t-shirt's label. Actually

Threadless has over one million registered users and receives approximately 800 submissions per week, six of these are offered a week.

This method of customer co-design, exploits the commitment of users to screen, evaluate and score new designs as a powerful mechanism to reduce flops of new products by empowering them to help. The method breaks with the known practices of new product development. It utilizes the capabilities of customers and users for the innovation process.[5] Together with just 20 employees, the company's founders sell more than fifty thousand t-shirts and earn profits amounting to over one hundred thousand dollars per month. This is achieved by transferring all essential productive tasks to their customers who, in turn, fulfill their part with great enthusiasm. Customers design their own t-shirts and help improve the ideas of their peers. They screen and evaluate potential designs, selecting only those that should go into production. Since customers (morally) commit themselves to purchase a favored design before it goes into production, they take over market risk as well. Customers assume responsibility for advertising, supply models and photographers for catalogues, and solicit new customers. The process starts when an idea for a product is posted on a dedicated web site by either a (potential) customer or the developers of a manufacturer. Second, reactions and evaluations of other consumers towards the posted idea are encouraged in form of internet forums and opinion polls. Based on the results of this process, the manufacturer investigates the possibility of commercialization of the most popular designs. Is this evaluation positive, the company decides about a minimum amount of purchasers necessary to produce the item for a given sales price, covering its initial development and manufacturing costs (and the desired margin). The new product idea is then presented to the customer community, and interested customers are invited to express their commitment to this idea by voting for the design or even placing an order. Accordingly, only if the number of interested purchasers exceeds the minimum necessary lot size, investments in final product development are made, merchandising is settled and sales are commenced.[6]

At Threadless, the entire business model is based on customer co-design. Users can evaluate each week between 400 and 600 new designs on a scale from zero ("I don't like this design") to five ("I love this design"). In average, each design is scored by 1500 people. A good score corresponds to a value above 3.0. But in addition, customers not only express their marked preference for specific designs, but can also opt-in to purchase the design directly once it has been chosen by the collective. For this, they check a box "I'd buy it" next to the scale. From the designs receiving the top votes and largest commitment of users to purchase, Threadless is producing today between four to six new products each week. To keep the competition interesting and encourage users to participate continuously, the number of designs at one give time has to be limited so that users don't get confused. Usually, each design gets seven days to be scored. But if a new design has received a low arbitrary score (made up of multiple variables including the number of "I'd buy it" requests and the design's average score) within the first 24 hours of its positing, it will be dropped from the running. This happens to about one third of the submissions. The early user feedback has proven to be a very strong indicator of the success of a design in the competition and enables the company to increase the usability and experience for users who vote. Motivated by its success in the fashion market, Threadless' founders have recently extended their categories to formal wear like ties or polo shirts (NakedandAngry.com) or music (15MegsofFame.com).[7]

3 Collective Customer Commitment versus Postponement and Mass Customization

Thus, manufacturers had to find new ways to increase the probability in meeting heterogeneous and fast changing customer needs. Studies have shown that the forecasting accuracy can be improved dramatically after observing just twenty percent of the initial sales of an item.[8] Companies have reacted on this insight by delaying some activities, rather than starting them with incomplete information input, to cope better with environmental uncertainty inherent to dynamic markets. In such a *postponement* strategy, manufacturing is split into two phases: in an initial phase, (generic) components are build-to-stock, and in a second stage, these components are transferred into the final product specification once more information about the market demand is available.[9] Connected with postponement, but different in nature, is *mass customization*. While in a postponement system the products are typically pre-defined by the supplier, with mass customization this process is reversed. It starts with customers co-designing their products, using a configuration system to specify their preferences. The individualized product is then manufactured on-demand.

Postponement and mass customization offer additional flexibility to minimize the new product development risk, but this flexibility does not come without costs. Both strategies require a redesign of the products and processes. This includes the creation of modular product family structures and often heavy investments in new flexible machinery equipment. For mass customization, also an elicitation system has to be in place to access the preferences of each individual customer and to transfer them into a precise product definition. On the operational level, postponement and mass customization imply costs of less efficient processing. As a result, mass customization and postponement are discussed broadly in the management literature, but rather few companies have implemented these strategies successfully today.[10]

Now compare Threadless' method to postponement and mass customization (*see Figure below*). Threadless has substituted conventional market research by deep continuous interactions with its customers. It does not ask its customers what they want to wear, but gives them a platform where they can express themselves and design these products. But most important and contrarily to earlier observations of customer or user driven innovation (see below), Threadless also transfers the decision process about what will be produced or not into the customer domain. Threadless provides its customer community the capability to organize themselves and collect consensus over the most favorable upcoming products. Therefore we call this method "collective customer commitment". Remember: Only if enough customers pledge to purchase a new product design, the design will be finalized and go into production. In this way, market research expenditures are turned into early sales.

Threadless also needs less flexibility in its manufacturing system. Instead of investing in highly flexible manufacturing systems and dealing with individual custom designs, the company focuses its energy to motivate creative designers to submit new designs and facilitates the evaluation and voting process in its customer community.

Contrarily to postponement, it only starts the *full* manufacturing cycle after customers have shown their real commitment to purchase a particular item, eliminating the risk of product flops while allowing still for economies of scale. It also has not to make risky decisions about pre-fabrication or the optimal point of postponement. Compared to mass customization, Threadless has not to interact with individual customers and to run manufacturing lots of one. The costly elicitation process is substituted by an early involvement of some (expert) customers in development and the refinement of their ideas and pre-order taking by a larger group of customers. Likewise from the customers' perspective, the effort and risk to decide about a custom design – mandatory in a mass customization configurator – is replaced by the security of peer-evaluated products.

Postponement Strategy	**Mass Customization**	**Collective Customer Commitment Method**
new product development by manufacturer (based on market research input)	development of product architecture and customization options by manufacturer	development of new product design by some (expert) customers
▼	▼	▼
prefabrication of (some) components	customer co-design process (elicitation)	evaluation and refinement of design by manufacturer *and* customer community
▼	▼	▼
access to better market information (based on market research input)	placing of order by each individual customer	presentation of selected design concepts and obtaining commitment of potential customers
▼	▼	▼
final assembly of product variant	custom (on-demand) manufacturing	only if minimum lot size is pre-sold, (mass) production of product starts
▼	▼	▼
mass distribution	custom distribution	mass distribution

Fig. 1. The collective customer commitment method combines ideas of postponement and mass customization, but adds own characteristics as well

4 When Deep Integration of the Customer Makes Sense

Integrating customers in the innovation process and collecting customer purchase orders in advance of expenditures on detailed design and production: What may sound like an obscure idea of a small company in a niche market is becoming an increasingly popular approach with large companies as well. Indeed, in some markets this is the dominating way to make business: Consider the real estate market: here, condominiums are often sold like Treadless t-shirts: The developer will only start the construction when a given number of buyers have shown their willingness to purchase an apartment by placing a down payment. But what has been an approach for very costly products like condos in the past is passing downwards to fast-moving consumer commodities. We see two situations when the collective customer commitment method provides most value: (1) to test really innovative products where little customer experience exists and thus market research is very fuzzy, and (2) to create fitting products for rather small and very heterogeneous market segments.

Yamaha, a large manufacturer of musical instruments, employed the collective customer commitment method in the first situation. Yamaha's design team had envisioned an innovative electronic guitar, based on the feedback of frustrated, but lazy hobby musicians who wanted to play an instrument just without practice. The team came up with an instrument where, once fed with a song, small lights would tell the user where to press the fingers. This idea was breaking with the traditional design of a guitar and was considered too risky to be produced and developed in the conventional system. Thus, Yamaha used an existing user community to find out if there would be enough customer commitment for this design.[11] Users quickly draw on the idea and provided suggestions for improvements (like adding an amplifier and making the device battery-powered). Once the final design was posted by Yamaha, the minimum order quantity was reached almost immediately, motivating Yamaha to produce this product. Until today, it sold more than 20,000 units, five times more than the average product in this category.

The second situation relates to a market where customer demand is very heterogeneous, a common situation today in many markets due to fast changing trends and more diverse needs.[12] Also the borders of formerly local markets are diminishing, and customer needs become geographically broadly distributed. In heterogeneous and distributed markets, however, information about the demand for (new) products is distributed in an extremely diverse way, leading to large information asymmetries between individual customers and manufacturers. For manufacturers who want to provide an offering fitting exactly into such a market segment in order to exploit this differentiation opportunity promising high margins, it will become very costly to access all required information.[13] If the knowledge of manufacturers about the needs of an emerging market is scare and costly to achieve via conventional market research, user contributions are becoming a valuable source of innovation. The possibility of open contributions encourages a self-screening by potential contributors.

Research on customer or user innovators has identified that in many markets users with so called lead user characteristics exist.[14] These users realize a need for a new product (or functionality) ahead of the average users, or might be trendsetters or opinion leaders with regard to esthetic attributes. In our work with companies we often found that these customers are willing to disclose new ideas openly to the manufacturer and other users. They expect that their contribution will be of interest for others who will adopt the idea, develop it further, and make the resulting product cheaper when a manufacturer can produce the good for a larger group instead customized for just one client.

In such a situation – a specific new need is distributed highly heterogeneously among a large population of geographically spread customers – customers benefit from (i) becoming active by their own and develop and explore own ideas to fulfill a specific desire or need, and (ii) from organizing themselves as a group of users with similar needs in terms of the said product idea. While for high-involvement products customers may organize and foster this process by their own (consider patient groups who initiate, organize, and fund new research for new pharmaceuticals[15]), many users lack the motivation to transfer their need into a new product by themselves, but rely on manufacturers to do so. But a manufacturer has to be confident that a feasible demand for the proposed new product exists. He could try to investigate this demand by conducting costly and risky market research, but could also facilitate this group and organize the generation of collective commitment. This allows the manufacturer to profit from first-hand secure information about the scale of this need. He gets a first-mover advantage to step ahead with producing this product and harvest the new market segment. Instead of generating market research expenditures, collecting early customer commitment generates instant sales. The capabilities of online interaction via the internet enable this process today for almost all product categories, independently of their overall market value. Thus, the strategy of powerful real estate developers to hedge their risk by pre-selling apartments can now be repeated for almost every product and by every manufacturer.

The collective customer commitment method further recognizes that not everyone wants to actively participate in product development activities. Not all customers are lead users. Customers can decide about the degree of their involvement: At Threadless, most new designs are submitted by young professional designers, i.e. users with typical lead user or trendsetting characteristics. They contribute not only because the monetary incentive of $1000 is higher than the average honorarium paid for a commissioned design by a conventional clothing company (about $300 to $500). Their main motivator is to get larger exposure in the professional design scene, a rather closed market which is difficult to enter for newcomers. The openness of Threadless' community makes it easy for designers to present their work and to get immediate feedback. But Threadless allows also pure hobbyists to submit a design as the screening activities by its community enable this openness at no risk and with no costs. Others users just comment on the submissions and propose amendments or additions. The majority of Threadless' users, however, just screens the proposals and contributes to the elicitation of demand by polling for the designs they like most. For these customers, browsing through the ideas is often a novel experience and a welcomed change from traditional shopping activities.[16] They discover new potential products, exchange comments, and feel empowered by their authority to make a favorite idea happen.

5 Implementing the Collective Customer Commitment Method

Collecting customers' commitment and taking pre-orders before production starts is not new. This has been a common pattern in specialized industrial markets where a customized solution is produced for a specific buyer. Also real estate developers work according to this scheme, starting a new building development only after a specific numbers of units have been sold in a pre-signing phase. But what is new is that gathering collective customer commitment is becoming a much larger phenomenon, being applied on fast-moving consumer goods. What is also new is the strong integration of consumers not only in the evaluation of a product idea, but their intense participation in the design process itself. There are several benefits for manufacturers to implement the collective customer commitment method in such a way. By creating an open line for their customers, manufacturers get access to ideas for new products or even complete designs. Especially in markets targeting rather specialized segments or in very volatile markets influenced by fast moving fashion trends, supporting recent and potential customers to organize themselves as a group and to express commitment for a specific design turns market research expenditures into sales. Once this commitment is explicit, manufacturers can exploit this collective demand and serve the market very efficiently without the conventional costs of identifying this segment and the risk of developing and producing a not appealing offering.

An important condition to make collective customer commitment a success is the full disclosure of the entire process from initial consumer comments to final product commercialization. Often designers develop their products in secrecy, fearful of the prying eyes of competitors, for an ideal customer who may not actually exist. The collective customer commitment method builds on the integration of customers in an open innovation process. If development process is kept confidential, it is impossible to synchronize the activities of the developer and the consumers. For example, potential customers have to obtain a virtual picture of the prototype as early in the design process as possible so that both the developers and the users have the same mental picture of the concept. This demands an open, transparent development process contrarily to the conventional practice of keeping innovation closed and secret. From our interviews with designers and management of the companies practicing the method we learned that switching from a closed to an open mode is often difficult and requires sincere change management activities. To master this mental change is one of the largest success factors when a firm wants to profit from collective customer commitment.

But it is important to note that in the end management keeps the final word. Threadless learned that the collective input of their customers has to be combined with the companies' internal market knowledge to succeed successfully with the commercialization of the selected products. At Threadless, the winning designs are chosen from the top scoring designs, but they are not necessarily the top scoring designs. Important factors are the originality of the design (is it somehow timeless, not too similar to other recent winners), legal issues (are there any copyright related issues), and assortment policy (will the design contribute to a wide assortment of products).[17] For this decision process however the community provides again important input: The often long list of user comments about each design provides helpful information if a design is plagiarism, but also if it could be modified to look better.

Conventional product development and the collective customer commitment method thus have to be seen as supplementary – not as substitutes. Successful innovation management is like any other management task, first of all, a decision about trade-offs, choosing what to do and what not to do. There will be contingency factors in favor of a manufacturer-dominated innovation process without any participation of the customer. But there is no doubt that customer integration matters in the new product development process. We believe that collective customer commitment holds plenty of opportunities for companies to reduce the risks of new product development and overcome the obstacles of conventional market research (*see Box: Why we expect more companies to be using the collective customer commitment method*).

Manufacturers who want to utilize these benefits have to decide about several building blocks of the collective customer commitment method. They express alternatives to what extent a company wants to substitute conventional market research and product evaluation measures by customer participation (*see Box: The building blocks of the collective customer commitment method*). We expect that promising fields to apply the collective customer commitment method include fashion items, household utensils, sports goods, home appliances and consumer electronics, but also the development of future prefabricated houses, automotives or machinery of specialized applications. The beauty of the method is that exploring it does not come at much cost: If no customers opt-in to give their commitment for one particular design, the company has not lost much. This experience, even if it may be disappointing, comes much cheaper than producing and distributing high volumes of products which in the end no one wants – quite a familiar situation for many product managers today.

Notes:

[1] Balachandra & Friar (1997); Urban & Hauser (1993); Poolton & Barclay (1998); Redmond (1995); Tollin (2002).
[2] Henkel & von Hippel (2005). Refer also to Adams et al. (1998); Bacon et al. (1994); Teas (1994).
[3] Burke (1996) provides a good review of the inefficiencies of traditional market research.
[4] Adams et al. (1998); Mahajan & Wind (1992).
[5] A good review of research on customers as sources of innovation provides von Hippel (2005). Sawhney, Prandelli and Verona (2003) show that these customers are often organized in communities by a manufacturer or intermediary. Piller et al. (2005) comment on the opportunities to perform co-design activities in a community.
[6] The origins of the idea can be traced back to Kohei Nishiyama and Yosuke Masumoto, two industrial designers from Tokyo. In the 1990s, they pioneered the idea with their company Elephant Design. The core element of the company is its website cuusoo.com (cuusoo means "ideal" or "daydream" in Japanese). Here consumers can post ideas for desired products. One idea, for example, came from a copyeditor who used his home as an office and wanted a discreet microwave, a plain white box. This seems to be an odd request, but when the company showed a virtual prototype, many users expressed consent. In the academic literature, Elofson and Robinson (1998) describe a similar system called "custom mass production": Users first negotiate on a particular product design, find consensus about a solution that is fitting the desires of all, and auction the resulting common to interested manufacturers.
[7] A company with a very similar business model is Buutvrij from The Netherlands (www.buutvrij.com).
[8] Fisher & Raman (2001).

[9] McCutcheon, Raturi & Meredith (1994).
[10] See with regard to postponement Gupta & Benjaafar (2004); Skipworth & Harrison (2004); with regard to customization Agrawal et al. (2001); Zipkin (2001).
[11] Yamaha teamed up with Engine, Inc., a competitor of Elephant Design (see note 6). Engine focuses on fashion items and the merchandizing of movie and comic characters (its 2004 sales topped 570 Million Yen). Registered users can submit "please, make this" posts, i.e. ideas for new products, on its web site tanomi.com (the name derives from the Japanese term tanomikomu, meaning requesting, referring both to the consumers' requests to produce a design and the manufacturers' request to purchase the product before production). Once copyright and production feasibility are cleared by a company board, the idea is published to the whole community for evaluation, together with a price and minimum order quantity for its commercialization. In addition, Engine offers other manufacturers to post innovative product concepts directly to its community.
[12] See Zuboff & Maxmin (2002) for an analysis of the reasons why markets are becoming more heterogeneous.
[13] Von Hippel (2005:72-75) calls these domains where large information asymmetries between individual users and manufacturers exists "low-cost innovation niches", i.e. fields where information held locally by individual users strongly motivates them to contribute actively to a new development. With regard to this information transfer problem, see also von Hippel (1994) and Ogawa (1998).
[14] Von Hippel, Thomke & Sonnak (1999).
[15] An example for such a patient group is ALS Association (also.org). Here, patients with Amyotrophic Lateral Sclerosis commission own research to find treatments for their disease.
[16] On the internet, a growing number of websites serves this demand of innovation-seeking consumers (e.g., gizmodo.com, coolhunting.com or boingboing.net). They allow, however, only discovering existing new products, but do not provide any open line to the manufacturers or product developers.
[17] The Threadless team also goes through each short listed design to make sure there was not any cheating involved by analyzing IP addresses and IP chains for voters and the respective scores given.

References

Adams, M.E., Day, G.S., Dougherty, D.: Enhancing New Product Development Performance: An Organizational Learning Perspective. Journal of Product Innovation Management 15, 403–422 (1998)

Agrawal, M., Kumaresh, T.V., Mercer, G.: The False Promise of Mass Customization. McKinsey Quarterly 38(3), 62–71 (2001)

Bacon, G., Beckman, S., Mowery, D., Wilson, E.: Managing Product Definition in High-Technology Industries. California Management Review 36, 32–56 (Spring 1994)

Balachandra, R., Friar, J.H.: Factors for Success in R&D Projects and New Product Introduction. IEEE Transactions on Engineering Management 44(3), 276–287 (1997)

Burke, R.: Virtual Shopping: Breakthrough in Marketing Research. Harvard Business Review 74, 120–129 (1996)

Danneels, E.: Tight-Loose Coupling with Customers: The Enactment of Customer Orientation. Strategic Management Journal 24, 559–576 (2003)
Eisenhardt, K.M.: Building Theories from Case Study Research. Academy of Management Review 14(4), 532–550 (1989)
Elofson, G., Robinson, W.N.: Creating a Custom Mass-Production Channel on the Internet. Communications of the ACM 41, 56–62 (1998)
Fisher, M., Raman, A.: Reducing the Cost of Demand Uncertainty Through Accurate Response to Early Sales. Operations Research 44, 87–99 (2001)
Gupta, D., Benjaafar, S.: Make-to-Order, Make-to-Stock, or Delay Product Differentiation? A Common Framework for Modeling and Analysis," IIE Transactions 36, 529–546 (2004)
Gummesson, E.: Qualitative Methods in Management Research, 2nd edn. Sage, Thousand Oaks (2000)
Henkel, J., von Hippel, E.: Welfare Implications of User Innovation. Journal of Technology Transfer 30, 73–88 (2005)
Mahajan, V., Wind, J.: New Product Models: Practices, Shortcomings and Desired Improvements. Journal of Product Innovation Management 9, 128–139 (1992)
McCutcheon, D.M., Raturi, A., Meredith, J.R.: The Customization-Responsiveness Squeeze. Sloan Management Review 35, 89–99 (Winter 1994)
Ogawa, S.: Does Sticky Information Affect the Locus of Innovation? Evidence from the Japanese Convenience Store Industry. Research Policy 26, 777–790 (1998)
Piller, F., Schubert, P., Koch, M., Moeslein, K.: Overcoming Mass Confusion: Collaborative Customer Co-Design in Online Communities. Journal of Computer-Mediated Communication 10(4) (2005), http://jcmc.indiana.edu/vol10/issue4/piller.html
Poolton, J., Barclay, I.: New Product Development from Past Research to Future Applications. Industrial Marketing Management 27(3), 197–212 (1998)
Redmond, W.H.: An Ecological Perspective on New Product Failure: The effects of Competitive Overcrowding. Journal of Product Innovation Management 12, 200–213 (1995)
Reinmoeller, P.: Dynamic Contexts for Innovation Strategy: Utilizing Customer Knowledge. Design Management Journal Academic Review 2(1), 37–50 (2002)
Sawhney, M., Prandelli, E., Verona, G.: The Power of Innomediation. Sloan Management Review 44, 77–82 (Winter 2003)
Skipworth, H., Harrison, A.: Implications of Form Postponement to Manufacturing: A Case Study. International Journal of Production Research 42(10), 2063–2081 (2004)
Teas, R.K.: Expectations as a Comparison Standard in Measuring Service Quality: An Assessment of a Reassessment. Journal of Marketing 58, 132–139 (1994)
Tollin, K.: Customization as a Business Strategy: A Barrier to Customer Integration in Product Development. Total Quality Management 13, 427–439 (2002)
Urban, G., Hauser, J.: Design and marketing of new products, 2nd edn. Prentice Hall, Englewood Cliffs (1993)
von Hippel, E.: Democratizing Innovation. The MIT Press, Cambridge (2005)
von Hippel, E.: Sticky Information and the Locus of Problem Solving. Management Science 40, 429–439 (1994)
von Hippel, E., Thomke, S., Sonnak, M.: Creating Breakthroughs at 3M. Harvard Business Review 77, 47–57 (1999)
Zipkin, P.: The Limits of Mass Customization. Sloan Management Review 42, 81–87 (2001)
Zuboff, S., Maxmin, J.: The Support Economy: Why Corporations are Failing Individuals and the Next Episode of Capitalism. Viking Penguin, London (2002)

Author Index

Abbasi, Alireza 49
Abramovich, A. 121
Altmann, Jörn 49, 93

Busch, Michael W. 107

Fischbach, Kai 75, 135
Fliess, Sabine 13
Floeck, Fabian 75
Fuehres, Hauke 135

Georgi, Sandro 63
Gloor, Peter A. 135
Guo, P. 121

Hwang, Junseok 93

Ickler, Henrik 25

Jung, Reinhard 63

Kim, Kibae 93
Krauss, Jonas 135

Lindgens, Evalotte 147

McGovern, Mark 1

Nadzeika, Arwed 13
Nann, Stefan 135

Piller, Frank T. 147
Putzke, Johannes 75

Qian, T. 121

Schoder, Detlef 75
Sheu, P.C.-Y. 121
Steinfels, Sabrina 75

Tacke, Oliver 37

von der Oelsnitz, Dietrich 107

Wang, L. 121
Wang, Q. 121
Wehler, Marco 13
Wormsbecher, Jorinde 13

Xu, C.Z. 121